GRASSHOPPERS AND MANTIDS OF THE WORLD

GRASSHOPPERS AND MANTIDS OF THE WORLD

Ken Preston-Mafham

Photography by Ken and Rod Preston-Mafham
of Premaphotos Wildlife

BLANDFORD

Paperback edition first published in the UK 1992
by Blandford, a Cassell imprint

Cassell plc,
Wellington House
125 Strand
London
WC2R OBB

Reprinted 1998 & 1999

Previously published in hardback by Blandford in 1990

Distributed in the United States by
Sterling Publishing Co., Inc.,
387 Park Avenue South, New York, NY 10016–8810

A Cataloguing-in-Publication Data entry for this title is available
from the British Library

ISBN 0–7137–2381–5

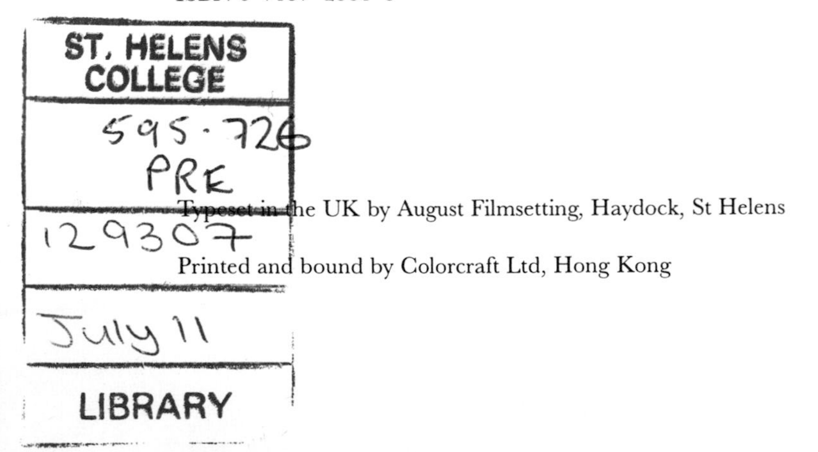

Typeset in the UK by August Filmsetting, Haydock, St Helens

Printed and bound by Colorcraft Ltd, Hong Kong

Contents

Acknowledgements

The photographs are the result of visits to 24 countries around the world, mainly in the tropics. The success of these trips has been greatly assisted by a number of generous people, without whose help I would certainly not have been in a position to devote so much time to finding and photographing my subjects.

My special thanks go to the following: Dr Victor Becker, Brazil; Adriana Hoffmann, Chile; Dr Mario Boza, Costa Rica; Dr Angus McCrae and his wife Janet, Kenya; Ken Scriven, Malaysia; Dr Roger Kitching, Australia; Dr John Ismay, New Guinea; Dr Amnon Friedberg, Israel; Patrick Daniels of Duke University, Madagascar.

I have also been fortunate to enjoy the assistance of a number of specialists on the orthopteroid insects. Dr C. Amédégnato in Paris identified most of my Peruvian grasshoppers, while Mrs Judith Marshall of the Natural History Museum in London deserves my special thanks for much help and encouragement over the past ten years. Linda Pitkin, also of the Natural History Museum, helped with many Kenyan specimens.

Finally, I would like to thank Alan Rollason for the line drawings which appear in the book, and my brother, Rod, and his wife, Jean, for looking after the book during the various production stages, while I was abroad.

K. P -M.

Preface

For many years I have felt that the grasshoppers and mantids deserve to be made known to a wider public. I am sure that few naturalists familiar only with, say, the limited northern European selection of generally small, drab grasshoppers would be aware that some of the most spectacularly coloured insects in the world belong to the very same clan. It has therefore been with considerable pleasure that I have been able to sit down and write the present volume. I have been more privileged than most to roam the world in search of natural subjects in my profession as a wildlife photographer. These travels have brought me into contact with a wide variety of insects, many of which are far more colourful than their British counterparts, which formed my first studies. In no group was this difference more notable than in the grasshoppers, while mantids are, alas, completely absent from my home country. The same remarks apply, to a lesser extent, to the grasshoppers over much of the United States, where some very beautiful species do indeed occur, but can only be seen by those willing to search the right places in the deserts of the southwest. A book which can communicate in words and pictures some of the beauty and fascination of these insects to a far wider audience is therefore long overdue.

The term 'orthopteroid', which is used throughout this book, requires some explanation. Many years ago, quite a wide range of insects having biting mouthparts were included in a single large order, the Orthoptera. Over the years this has been split up into a number of small orders, not all of which are dealt with in the present work. The termites (order Isoptera), earwigs (order Dermaptera) and members of the order Grylloblattodea were all once in the Orthoptera but are not mentioned in the text, which concentrates on the cockroaches (order Blattodea), praying mantids (order Mantodea), grasshoppers, katydids, crickets and mole-crickets (order Orthoptera as now accepted) and stick insects or walking sticks (order Phasmatodea). These are the 'orthopteroid insects' mentioned throughout the text.

The common names of insects differ from one country to another, even when the same language may be common to both. Thus in Britain members of the Tettigoniidae are called bush-crickets. In the United States these insects are called katydids (a word derived from the call of one or two species), while a bush cricket is a true cricket and not a tettigoniid. Even without the hyphen, this is so similar to the British term that I have felt it better to use the American word katydid throughout the book. However, 'stick insect' and 'walking stick', as used in Britain and the USA respectively, are both readily understandable and I have used the British term throughout.

I have kept the use of technical jargon to a minimum; along with other titles in this series, the book is intended to be used by the curious layman as well as by the experienced biologist. For the same reason, details of the physiology and internal workings of the subjects are kept to a basic minimum; such information is more relevant to the specialist than the amateur and is readily available in numerous textbooks specializing in insect physiology. Instead, I have devoted the maximum amount of space to describing the main aspects of behaviour, hoping that this will stimulate the reader to make original observations on his or her own account. The breadth of the coverage allocated to the various orders and families tends to reflect the relative numbers of species, which itself dictates the frequency with which the insects may be encountered and therefore the opportunities to take an interest in them.

The photographs are all drawn from our picture library Premaphotos Wildlife and were selected from our coverage of just over 3000 pictures of orthopteroid insects depicting more than 350 different species. All the photographs were taken in the wild, during visits to some 24 countries, and were *not* artificially posed for photographic purposes. Even the most camouflaged of subjects was detected by the author the hard way – by painstaking, lengthy searching in the natural habitats, particularly tropical rainforest, rather than by collecting with a net or using a light-attractant at night. Only by finding the insects actually resting up during the day can there be any guarantee that the background on which the subject is posed has been selected by the insect itself and not by the photographer seeking to create a saleable picture, as so often happens with the photography of camouflaged insects.

Ken Preston-Mafham
King's Coughton
England

Chapter 1

Classification and Physiology

It is difficult to think of any group of living organisms whose system of classification is accepted by all the workers specializing in the group concerned. The insects dealt with in this book are no exception. At one time they were all included under the broad heading of the Orthoptera, but the splitting up of this very large and heterogeneous grouping into a number of smaller orders seems to have won broad acceptance. What is less widely accepted is the recent promotion to full family status of many former subfamilies falling within the current more restricted concept of the order Orthoptera. This splitting has given rise to large numbers of 'families', many of which include only a handful of species. The author feels that the taxonomic treatment of the insects as a group should at least try to be reasonably consistent from one order to another. If entomologists dealing with the order Hymenoptera can accept 14,000 (and in reality probably far more than double that figure) species of ants in the single family Formicidae, and workers on the Coleoptera (beetles) can live with 35,000 or more different weevils in the family Curculionidae, then there seems little justification for dividing the 5000 strong Tettigoniidae into 11 families and the 10,000-plus members of the Acrididae into 12 families. The members of the Acrididae are all variations on a single, fairly uniform theme, with relatively minor differences in physiology and behaviour to suit divergence in lifestyles and habitats. The characters used to justify erection to full family status would probably not be accepted as being of sufficient importance for such purposes in many other groups of insects. For this reason, and bearing in mind that taxonomy in general seems to be becoming more conservative in outlook (e.g. the fairly recent reduction of the butterfly families Heliconiidae, Ithomiidae, Acraeidae, Amethusiidae, Morphidae, Libytheidae and Brassolidae to subfamilies within the Nymphalidae), the author has chosen to adopt what some may consider to be a rather old-fashioned and very conservative scheme of classification in the present work.

ORDER BLATTODEA

After their initial separation from the Orthoptera, the cockroaches were at first included along with the praying mantids in the order Dictyoptera. Both groups, however, are now included in their own distinct orders, a move which seems to have won acclaim from most entomologists. Cockroaches are a very ancient group with a fossil record dating back over 300 million years. Just over 3500 species have been described, although a similar number is probably still awaiting discovery and description. Most species are found in the warmer parts of the world, especially the tropics.

Cockroaches are often used as examples of 'typical' insects of the type

Plate 1 *Elliptorhina javanica* is a large, wingless cockroach from Madagascar. It hisses when molested and emerges at night to feed on fallen fruits and other plant debris, hiding away during the day in hollow logs.

having chewing mouthparts, and will serve here to describe the special characteristics of an insect's makeup. We can take that common pest, the large American cockroach *Periplaneta americana*, as our model. The body is split into three distinct regions: the head, the thorax and the abdomen. The head bears a pair of antennae, the eyes and the mouthparts. The antennae are the cockroach's 'nose', for they are covered in vast numbers of minute pits and sensory hairs which are incredibly sensitive organs of smell and touch. The mouthparts consist of two pairs of jaws, the hard jaws or mandibles and the soft jaws or maxillae positioned between an upper lip or labrum and a lower lip or labium. When a cockroach (or grasshopper etc.) chews its food, the jaws operate from side to side rather than up and down as in humans and other vertebrates. The cockroach's compound eyes are even more different, consisting of large numbers of tiny facets which given the eye the appearance of a honeycomb when seen under a lens.

The thorax is extended forwards to provide a protective shield for the head and contains three segments, each sporting a pair of legs. The leg is made up of five parts: the coxa, a short thick connection between leg and thorax; the trochanter, a short joint connecting the coxa to the next part, the femur, followed by the tibia and finally the foot or tarsus formed of five small joints and finishing in a pair of claws. Many cockroaches have relatively long legs and can run extremely fast to escape from enemies. The American cockroach *Periplaneta americana* can manage a short sprint at 1.65 mph (2.65 kph), although it soon tires. Compare this, though, with the 0.036 mph (0.058 kph) of a black ant *Lasius niger* at full tilt. Not all cockroaches have wings, but in those that do they consist of two pairs positioned on the second and third segments of the thorax. The first pair of wings is leathery and acts as protective covers for the second pair, which is far more flexible and does most of the work during flight. The abdomen is split into ten segments and is tipped by a pair of slender feelers called cerci.

The cockroach, like all insects, wears its skeleton on the outside, the exo-skeleton, which is made of a tough, highly durable material called chitin. This is normally quite hard, so in order to grow insects have to undergo a series of moults, during which each new exo-skeleton is soft but durable, and the insect is able to expand to fit the new larger size. Along the sides of the body are ten small openings or 'spiracles' which admit air into the body cavity via a pumping action of the abdomen. The air is carried to various parts of the body not via the blood but in a series of branching thread-like tubes called tracheae. The heart is situated along the back and consists of a tube open at the front end and closed at the rear. Expansion of the heart draws in blood through a series of one-way valves situated on the sides. Contraction squirts the blood out of the front end.

The female reproductive organs are composed of eight egg-tubes on either side. Towards the rear these unite to form the left and right oviducts which then link up to form the vagina. Sixteen eggs mature at a time, gradually growing as they move along the egg-tube. As they are laid, the eggs are enveloped in a protective froth produced by a pair of glands opening into the vagina. The male cockroach has a pair of small testes which open to the exterior via an ejaculatory duct. At its front end this duct has a large mushroom-shaped gland which produces a disk-like blob called the spermatophore, which acts as the carrier for the male cells during copulation. In some

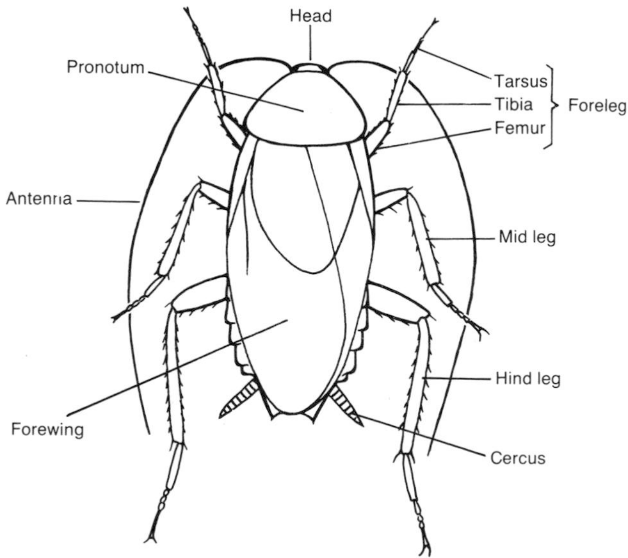

Fig.1 A typical cockroach *Periplaneta australasiae*, a pest species.

katydids in the order Orthoptera this spermatophore is enormous – as much as 40% of the male's total body weight – and its function is to provide the female with a nutritious meal after mating.

Five families are currently accepted within the Blattodea. The **Blattidae** contains about 40 genera and 520 species of medium-sized insects, including two familiar pests, the American cockroach *Periplaneta americana* and the Australian cockroach *P. australasiae*. The **Blattellidae** is the largest family, containing just over 200 genera and about 1750 species, some of which, e.g. species of *Megaloblatta*, are among the largest cockroaches, with a length of 4 in (10 cm). This family contains two of the worst cosmopolitan pests, the German cockroach *Blattella germanica* and the oriental cockroach *Blatta orientalis*. The **Blaberidae**, with 155 genera and just over 1000 species, is the second largest family and contains the most highly evolved species, the females often retaining the eggs inside the body until the young are ready to hatch. Some of the largest and most spectacular species resident in tropical rainforests and bat caves are found in this family. Members of the subfamily Panesthiinae feed on wood. The **Polyphagidae** has 39 genera and 190 species of often rather delicate, primitive coachroaches, some living in ants' nests. The **Cryptocercidae** contains only one genus with three species of primitive cockroaches which feed on wood. Two species come from China while *Cryptocercus punctulatus* is found in the United States.

ORDER MANTODEA

The name of this order is derived from the Greek word signifying a prophet or soothsayer. Though the predatory traditions of the praying mantids would scarcely seem to earn them such a designation, their habit of posing with their front legs held up before the face, as if in prayer, has given rise both to this and many of the common names in use around the world. The Mantodea is a relatively small order containing only around 1800 species in eight families. The vast majority are found in the tropics, with Africa boasting the largest variety of species, 880 contained in 156 genera (although 54 of these are only found in Madagascar). Asia comes next with 530 species in 144 genera, followed by the Americas with 410 species in 96 genera, few of which reach the United States. Oceania boasts 165 species in 50 genera while Europe can only muster 24 species in 13 genera. The smallest mantis, *Bolbe pygmaea*, is around $\frac{2}{5}$ in (1 cm) in length while the largest, such as some of the African *Tenodera* or *Archimantis latistyla* can reach 6 in (15 cm) or more.

Mantids are among the most easily recognized of all insects. The head is distinctly triangular with the large compound eyes set high up on either side. Three simple eyes or ocelli are normally present, although these are sometimes lost in males. This is an important distinction between mantids and cockroaches, which never have ocelli. The head is extraordinarily mobile with a very flexible articulation between the head and the prothorax, which is the name given to the extremely elongated first section of the thorax, whose topmost protective covering is called the pronotum. The front legs are long and heavily spined, forming efficient traps for the capture of living prey, the only food taken by mantids. Females are often wingless but males are usually fully

Plate 2 *Sphodromantis centralis* from Kenya is a fairly 'typical' green praying mantis. Clearly visible are the raptorial front legs with rows of prey-impaling spines, the triangular head with the large eyes mounted on the top corners, the long hair-like antennae, the elongate prothorax and wings folded lengthwise over the abdomen which bears two short slender cerci at its tip.

winged and fly to the females for mating. At rest the wings are folded lengthwise over the abdomen, which is divided into 11 segments and bears a pair of cerci near its tip.

The order is currently divided into eight families. The **Chaeteessidae** is neotropical in distribution and contains but a single genus, *Chaeteessa*, with a few species which are fully winged in both sexes; the front legs lack the highly developed spines found in other mantids. The **Metallyticidae** again only boasts a single genus *Metallyticus*, with a few species restricted to the Malayan region; compact insects with metallic bluish or green colours, a pronotum which is nearly square, and strongly spined front legs. The **Mantoididae** contains a single small genus *Mantoida*, delicate insects from the neotropical region. The **Eremiaphilidae** are thick-set wingless mantids with long slender legs suitable for running quickly across the sun-baked ground of their arid habitat. They are widespread across the desert and semi-desert regions of northern Africa and Asia, where their resemblance to stones makes them difficult to spot. The **Amorphoscelididae** are small mantids in two subfamilies, the Amorphoscelidinae from the Mediterranean and Oriental regions and Africa and the Paraoxypilinae from the sub-Saharan Africa and the Australian region. The **Hymenopodidae** contains some of the most spectacular mantids, some of which, such as the African *Pseudocreobotra ocellata*, are amazing mimics of flowers. The inner margins of the front femora are armed with a row of alternating long and short spines. The front wings are often

Plate 3 *Harpagomantis discolor* is one of several mantids in the subfamily Hymenopodidae which mimic flowers. The species is suitably coloured to blend in with a variety of different types of bloom. This specimen was one of several found in close proximity on savannah in the Orange Free State in South Africa.

decorated with bands or eye-like markings. The Hymenopodidae are found throughout the tropics except in Australia. The **Empusidae** are bizarre, rather slender creatures in which the antennae of the male are long and pectinate (i.e. with comb-like projections along both sides), rather than thin and bare as usually found in mantids. The running legs are often adorned with leaf-like lobes on the femora. This family includes eight genera with a rather small number of species scattered across Africa, the Mediterranean region and Asia. The **Mantidae** contains the vast majority of genera (263) and species within the Mantodea and not surprisingly shows a high degree of diversity in size and form. Most of the 'ordinary' green/brown or grey mantids belong to this family, as well as most of the species which mimic grass or bark. The male antennae may bear rows of cilia but are never doubly pectinate as in the previous family. The eyes are usually very large and the wings often reduced, at least in the female.

ORDER ORTHOPTERA

It is within the Orthoptera that such a complicated situation exists concerning how the order is to be classified into suborders, infraorders, superfamilies, families and subfamilies. Several schemes are available, but as none seems to agree completely, if at all, with any other, I am using a variation of one of the older schemes until such time as more people agree on something to replace it.

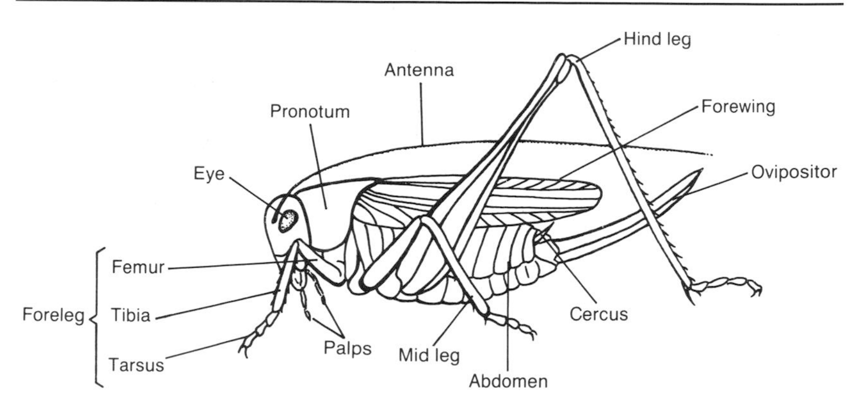

Fig.2 A typical katydid or bush cricket of the family Tettigoniidae.

The order is split into two suborders: the **Ensifera**, containing such familiar insects as crickets, mole-crickets and katydids, and the **Caelifera**, in which the grasshoppers are the primary and most well-known element. Each sub-order is further subdivided into two superfamilies and a relatively small number of families. The most characteristic aspect of the Orthoptera is the hind legs, which are usually long and strongly modified for jumping. Some crickets and grasshoppers are impressive performers and can leap 8 or 10 ft (2.3 or 2.6 m), being able to repeat the performance many times in succession without showing signs of tiring. This is possible because the back legs contain a protein called resilin, which has superb elastic properties with an incredible 97% efficiency in returning stored energy. The explosive release of this energy catapults the grasshopper instantly into the air, something which would not be possible if mere muscle-power were employed.

Before considering the various groupings, it is worth looking at some of the major differences between members of the two suborders, some of which are easily visible and will permit instant recognition, even in the field. Most members of the suborder Ensifera are easily separated from those in suborder Caelifera on details of the antennae. These are usually very hair-like and much longer than the body (sometimes several times as long) in the Ensifera (though not in the mole-crickets); usually shorter than the body and thicker and not at all hair-like in the Caelifera. Female Ensifera are notable for the long, sword-like egg-laying instrument or ovipositor which usually protrudes conspicuously from the rear end of the body. This is always absent in the Caelifera. The only possibilities for doubt may lie in separating a green grasshopper from a green katydid. Females can be separated immediately merely by the presence or absence of an ovipositor, but males will need an inspection of the antennae to make an instant judgement.

Another major difference lies in the method of sound production and reception typical of the two suborders. In the Ensifera, e.g. katydids, a vein on the left forewing is modified and provided with teeth to form a multi-notched ridge (the file), which is rubbed against a hardened area (the scraper) on the rear edge of the right forewing (although the file and scraper are mounted on the right and left wings respectively in the Gryllidae and Gryllotalpidae). The

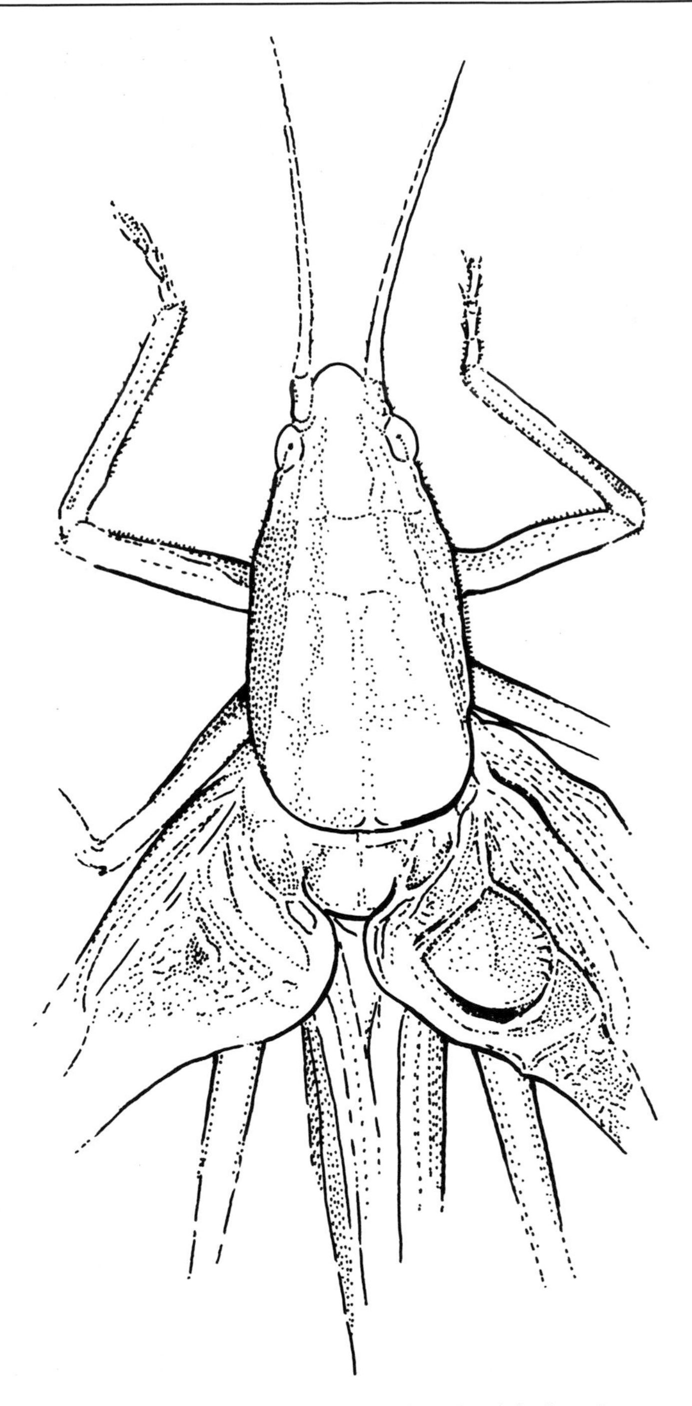

Fig.3 An African katydid showing the 'mirror' on the right forewing.

general efficiency of sound production is increased by the provision of a 'mirror', a small area of clear membrane situated at the base of the right forewing. In wingless species, the stridulatory apparatus is often the only remaining part of the forewings. The 'ears' are situated on the tibiae of the front legs, consisting of a small opening, the auditory slit, on each leg, leading to a pair of hearing membranes called tympana. In large tropical katydids this 'ear' can be easily seen with the naked eye. At the base of each front leg there is an aperture called the auditory spiracle connected to the tympana via a trachea which runs up the femur. The combination of trachea and spiracle gives the insect a very sophisticated omnidirectional hearing ability which alerts it to danger; while the auditory slits function as a unidirectional receiving system designed to guide the females accurately towards the sound beacon of a singing male.

The grasshoppers of the suborder Caelifera sing by scraping a row of minute pegs situated on the inner margins of the hind femora against the more prominent hardened veins of the forewings. The method of sound production in both the Ensifera and Caelifera is called stridulation. Some species amplify the sound by expanding parts of the forewing, while others inflate the whole abdomen to act as a resonating chamber. In some kinds the hind legs are merely flicked back on to the tips of the forewings to produce a 'ticking' sound. Many species are mute. The 'ears' are difficult to see, being situated on either side at the base of the abdomen, where they are often almost covered by the folded wings. The 'ear' consists of an external membrane or tympanum occupying a cavity which opens to the exterior via the tympanal aperture. It may be difficult to see the tympanum as the aperture is often partially blocked by an anterior flap.

SUBORDER ENSIFERA

SUPERFAMILY TETTIGONIOIDEA

FAMILY RAPHIDOPHORIDAE

This family includes the wingless camel crickets, the sand-treader crickets of North America and the cave crickets. These are generally rather drab, brownish insects, always wingless and without any obvious mechanisms for generating or perceiving sounds. The majority of species dwells in caves, although some dig small burrows in sandy deserts, while a few live in bushes. They are mainly scavengers, but will feed on butterflies and moths hibernating in caves, where the crickets' enormously long, thin antennae take over the task of locating prey and moving around, the eyes being of little use. The family includes around 300 species in 45 genera, spread around the world but seldom seen because of their retiring habits.

FAMILY STENOPELMATIDAE

Common names for these insects include stone crickets, New Zealand wetas, king crickets, sand crickets and Jerusalem crickets. They are usually very stout-bodied, brown, flightless, nocturnal scavengers, so are seldom seen except by the deliberate searcher. The head is usually very large and the

Plate 4 Members of the Stenopelmatidae are generally nocturnal, brown, wingless insects. This *Australostoma* species 'king cricket' in tropical rainforest in Queensland, Australia, has the enlarged jaws typical of the males in this family.

mandibles in the males may be grotesquely enlarged. The New Zealand wetapunga *Deinacrida heteracantha* is probably the world's heaviest insect. A large gravid female can measure $3\frac{1}{2}$ in (85 mm) by $1\frac{1}{4}$ in (32 mm) (excluding the ovipositor) and tip the scales at $2\frac{1}{2}$ oz (70 g). With their nocturnal foraging habits, catholic diet and large size, the wetas have been called 'invertebrate mice', having once occupied the ecological niche usually filled by rodents, of which New Zealand lacks any native species. Unfortunately rodents introduced by man have now almost totally ousted the giant wetas from their niche on all but a few offshore islands. Around 190 species in the family have been described from around the world.

Family Gryllacrididae

Popularly known as leaf-rolling crickets from their frequent habit of hiding during the day in a chamber made of a rolled leaf, these are drab insects in which the head is not noticeably outsize as in the preceding family and the wings are usually fully developed. They are mainly arboreal, preying on small, defenceless insects such as aphids or scavenging on dead invertebrates. Distribution is mainly tropical, with 550 species described in over 70 genera.

Family Cooloolidae

Although unlikely to be seen by any but the most ardent of Australian entomologists, the members of this family are so aberrant that the discovery of the first specimen rocked the entomological world. An insect discovered in south-eastern Queensland in 1976 was of such strange appearance that it was quickly

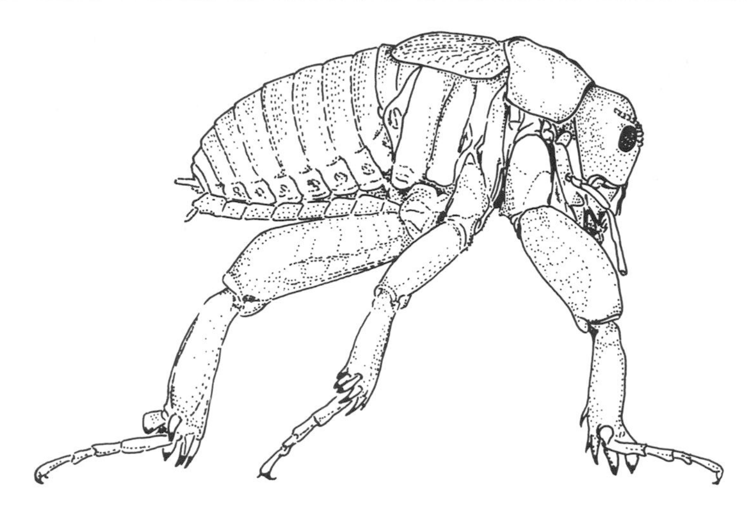

Fig.4 The Cooloola monster from Australia has its own family, the Cooloolidae.

dubbed the Cooloola monster. Powerfully built with a pale-coloured body, the new discovery was described as *Cooloola propator* and a new family was created to accommodate its peculiar characteristics. The most notable of these is the extremely shortened antennae which are uniquely reduced to only ten segments (the normal number in the Ensifera being 30 or more). The legs are stout and modified into shovels for digging in the ground, where the Cooloola monster seems to spend its whole life, preying on soil invertebrates, although the males may come to the surface at night after heavy rain. Two additional species have recently been found in Queensland, making three in the family.

FAMILY PROPHALANGOPSIDAE

This is a small family, with three species from North America and one from India (known only from a single specimen). The male stridulatory apparatus is primitive, although the sound produced is very loud. Mainly predatory and nocturnal and seldom seen.

FAMILY TETTIGONIIDAE

This is by far the largest family with around 5000 species in 900 genera, distributed worldwide but at its richest in the tropics. The method of stridulation has already been described above. The variety of body-forms is enormous, although broadly following subfamily lines (see below). Most species are green, although some are black, brown, grey or pink and a few have warning colours. Many species are fully winged and fly strongly to lights at night, when most species are active.

Some of the more interesting subfamilies, whose members may often be seen, or which figure prominently later in the book, are as follows. Members of the **Hetrodinae** (e.g. the South African armoured ground crickets or koringkrieks), **Bradyporinae** and **Ephippigerinae** (e.g. the familiar European 'tizi') are fat, slow-moving, flightless insects which live on or near the ground.

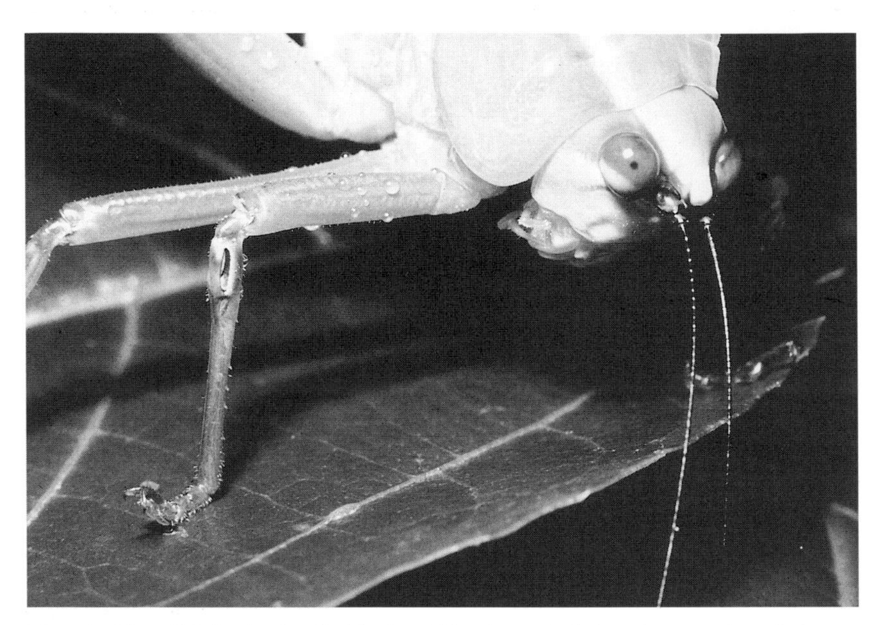

Plate 5 Katydids in the family Tettigoniidae have their 'ears' situated on their front legs. The auditory slit can clearly be seen on the tibia of this large *Microcentrum* species in tropical rainforest in Trinidad.

The males stridulate very noisily and have a large spermatophore. The antennae are unusually short and thick and may not exceed the body-length. The **Phaneropterinae** are mainly green, fully winged insects which often resemble leaves. The ovipositor is characterisically rather broad, flat and sickle-shaped. This is used to split the edge of leaves in which the eggs will be laid. The vegetarian **Pseudophyllinae** are moderately large to very large katydids, some with a wing-span of 8 in (20 cm). The majority are extraordinarily convincing mimics of leaves, either living or dead, and hail from tropical rainforests, where their song often rings loudly through the night air. Few penetrate into temperate areas, an exception being the true katydids *Pterophylla* from the eastern United States. The **Phyllophorinae** are the giants among the group and contain one of the world's largest and most spectacular insects, the magnificent 5 in (13 cm) long (excluding antennae) *Phyllophora grandis* from the primeval forests of New Guinea. The group as a whole is restricted to the Indo-Malayan-Papuan-Australian region. The shield-backed katydids of the **Decticinae** can easily be recognized by the large prothorax, which extends backwards over the first part of the abdomen. They are wingless, ground-dwelling insects with well-developed 'soles' on their hind tarsi. The **Copiphorinae** are sleek katydids with noticeably pointed heads, which in some neotropical species may be ornamented with a horn or 'crown of thorns'. They bite severely when handled and easily draw blood. Their diet is mixed. The African **Saginae** are exclusively carnivorous with strongly spined front legs.

Plate 6 Katydids of the subfamily Pseudophyllinae often mimic leaves. As its name suggests, this *Pseudophyllus hercules* from tropical rainforests in Malaysia is a particularly large and impressive species. The male's song rings out loudly through the forest at night.

Plate 7 Katydids in the subfamily Saginae are fierce carnivores with a painful bite and front legs densely armed with arrays of sharp spines. This is *Clonia wahlbergi* from mopane veldt in South Africa. The long ovipositor clearly indicates that this is a female.

SUPERFAMILY GRYLLOIDEA

FAMILY GRYLLOTALPIDAE

Mole-crickets are quite long, cylindrical, heavily built insects which could not be mistaken for any other orthopteroids. The antennae are very short for a member of the Ensifera, not surprising in insects which spend most of their lives in subterranean burrows. All the legs are modified for digging, particularly the front pair which is noticeably spade-like. Unlike most Orthoptera, mole-crickets cannot jump, as the hind legs are short and unsuited to the task. The wings are usually fully developed and mole-crickets may fly strongly, often whirring noisily into a lighted room at night. This can be particularly disconcerting in the tropics when a massed flight occurs. The front wings of the male bear stridulatory organs but no 'mirror'. The long ovipositor typical of the Tettigonioidea and Gryllidae is lacking. Only 60 or so species are known, grouped in a single genus *Gryllotalpa*, with a worldwide distribution, excluding the colder climates.

FAMILY GRYLLIDAE

A family of around 2000 species of mainly small to medium-sized, rather drab brown or black insects, often of retiring habits, sheltering by day under stones,

Plate 8 A typical *Gryllotalpa* mole-cricket of the family Gryllotalpidae. Note the antennae, which are much shorter than is general in the Ensifera, being far shorter than the body-length. Note also the spade-like front legs adapted for digging in the ground and the paired cerci at the tip of the body. Photographed in tropical rainforest in Trinidad during the day, having been flushed out of its burrow by rampaging army ants.

logs or bark, or in burrows, and emerging at night. Some species are diurnal and may be abundant and conspicuous insects. The song, however, is often loud and persistent, so crickets are typically heard rather than seen. The eggs are generally laid in soil. Most **Gryllinae** live on the ground, often spending at least part of their lives in burrows. The short-tailed crickets **Brachytrupinae** construct complex galleries and often constitute a serious pest of crops by damaging roots. The **Nemobiinae** or ground crickets have rather bristly bodies and often swarm in large numbers on the ground in suitable habitats; one European species is unusual in being intertidal. Members of the **Trigoniinae** lay their eggs inside plant tissues and prefer humid situations in rank vegetation near water. The hind tibiae of certain species are furnished with much enlarged spines which enable them to skate on the surface of water. The spider crickets and true cave crickets **Phalangopsinae** usually have very long slender legs and are mainly found in the American tropics, where most live among leaf-litter and rotting wood in rainforests. Caves are also favoured, while a few species are restricted to this habitat. The bush-crickets of the subfamilies **Eneopterinae** (Old World) and **Podoscirtinae** (New World) mainly live up off the ground in bushes and shrubs. The **Myrmecophilinae**

Plate 9 In the Ensifera, the antennae are often much longer than the body. This is taken to extremes in such species as this cricket *Phaephilacris spectrum* in the family Gryllidae in rainforest in Kenya. The lack of a noticeable ovipositor tells us that this is a male; note the long slim cerci at the tip of the abdomen.

or ant-loving crickets are the smallest of all orthopteroid insects, seldom over $\frac{1}{12}$ in (2 mm) and sometimes less than $\frac{1}{16}$ in (1.5 mm) long. The small eyes are reduced to only a few facets and the antennae are short. There are 45 species found more or less throughout the world, mostly in temperate or subtropical areas (but not in the British Isles). They are invariable found in association with ants or, occasionally, with termites and seem incapable of leading independent lives. Some species are ultra-conservative and live with only one species of ant; others are less fussy and will cohabit with a wide variety of hosts. Compared with the body-size the eggs are large and some species are habitually parthenogenetic, although this is probably not obligatory. The tree-crickets **Oecanthinae** are rather delicate, slender, pale-bodied insects seldom exceeding $\frac{1}{2}$ in (1.5 cm) in length. The head is positioned horizontally and the mouthparts jut forwards. The wings are always well developed; in males they are very broad and completely taken up by the enormous stridulatory organ with its particularly large 'mirror'. There are three genera and around 65 species distributed worldwide except in colder areas (e.g. there are none in the British Isles). The eggs are deposited within stems, twigs or bark, sometimes causing damage to orchard trees. Most species are predacious, especially on small helpless insects such as aphids, so here they have a beneficial role to play. The song is amazingly loud and melodious and the captive males are valued for their singing.

SUBORDER CAELIFERA

SUPERFAMILY ACRIDOIDEA

FAMILY PNEUMORIDAE

A small family with only 20 or so species, the 'bladder grasshoppers' are mainly found in the southern part of Africa. Most species are green, although a few are prettily marked with bands or spots of silver or red, and may be quite large, up to 4 in (10 cm) in length. In some species, the body of the male is semi-transparent, having become almost nothing more than an inflated balloon filled with air (hence another common name of 'flying gooseberries'). The body has, in effect, been transformed into a resonating chamber for amplifying the call, so the males could perhaps be considered as mobile loud-speakers. The stridulatory apparatus is unique to the family and diagnostic. The inner side of each rear leg has a miniature comb made up of a dozen or so rigid points. This scrapes against a series of raised dots or ridges situated on the sides of the abdomen. The hind legs in many species are mainly designed to act as fiddles and are poorly suited for jumping. The sound produced is amazingly loud for such a small creature and with its deep resonance is more reminiscent of large animals such as bullfrogs. The flightless females are quite different, with fat, non-hollow bodies.

FAMILY ACRIDIDAE

This is the main family of grasshoppers with more than 8000 species in 1500 genera grouped in a number of distinctive subfamilies. Distribution is worldwide and extends further into cold regions than is typical for the Ensifera. The

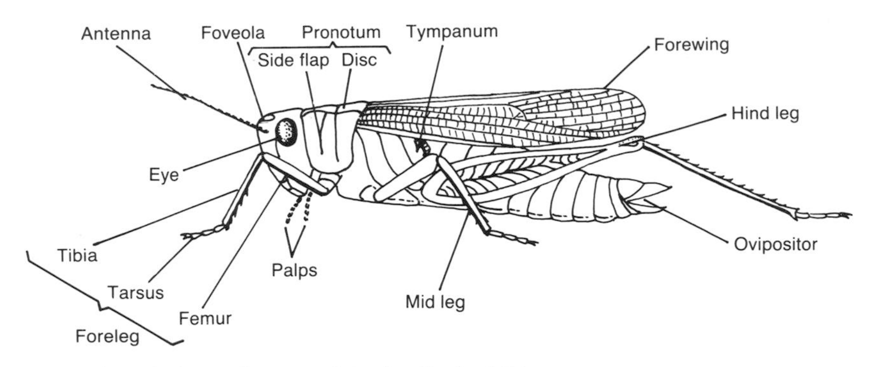

Fig.5 A typical grasshopper of the family Acrididae.

Plate 10 The common field grasshopper *Chorthippus brunneus* is a 'typical' temperate grasshopper, small and unimpressive. Note the antennae, much shorter than the body, a typical feature of most members of the Caelifera, and the enlarged back legs adapted for jumping. Temperate grasshoppers are experts at basking and this individual has drooped its rear leg downwards to expose the maximum amount of its abdomen to the sun, also tilting over slightly to one side against a reflecting yellow leaf. This specialized posture is regularly adopted by basking grasshoppers. Photographed in the British Isles where this is one of the most widespread and successful grasshoppers.

Plate 11 Members of the family Eumastacidae often have very long back legs which are held out at their sides. Note the typically short antennae on this nymph of a *Euschmidtia* species in tropical dry forest in Kenya.

Plate 12 'Monkey hoppers' of the family Eumastacidae usually have a noticeably high profile to the head and very short antennae. Many tropical species are brilliantly coloured. This gorgeous creature belongs to the genus *Eumastax*. It is a species new to science and was discovered by the author in rainforest near Satipo in Peru, an area which is rapidly being destroyed to make way for agriculture necessary for feeding an ever-expanding human population.

antennae are reasonably long (i.e. compared with the next two families, but much shorter than in most Ensifera), but only sometimes as long as the body. The range of sizes is enormous, from as little as $\frac{1}{3}$ in (1 cm) to the 6 in (15 cm) or more of an adult female *Tropidacris cristatus* with its 10 in (25 cm) wing-span and weight of 1 oz (30 g) or more. The sudden clatter made as this massive creature explodes unexpectedly into the air from a leaf beside one's ear and zooms across the sky is more reminiscent of a bird than an insect. Indeed, in Costa Rica, where this grasshopper is common, an ornithologist once shot one in mistake for a bird when collecting skins for a museum. The hind legs in most grasshoppers are well adapted for leaping, but their use in sound production probably applies to a minority of species, despite popular opinion to the contrary. Wings are frequently lacking and even fully winged species may be reluctant to fly for more than short distances, often only a few yards, using their wings more as sailplanes in a kind of glide-leap popular in the group. The hind legs of some species which spend almost their entire lives in water are adapted for swimming. Even the eggs are laid in floating water plants.

Three of the more interesting subfamilies are as follows. The **Pamphaginae** or toad grasshoppers are restricted to southwestern Asia and Africa, with 320 described species in 89 genera. These slow-moving rugose-bodied insects are among the most characteristic grasshoppers of the stony deserts of southern Africa such as in Namaqualand. The pronotum is usually heavily built and often has a raised crest. They are ground-dwelling insects, reluctant to fly and often resemble dead leaves, stones, earth or flaking bark. There is often strong sexual dimorphism with fully winged males and wingless females. There is a notably wide variety of calling mechanisms within this group. The **Pyrgomorphinae** contains some of the most colourful of all orthopteroids, perhaps of all insects. Members of this subfamily, which contains around 450 species, are often instantly recognizable by the conical shape of the head with a decidedly receding 'chin', although this is not always so. Most species live off the ground in trees and bushes and are cryptically coloured, although there are many examples of warningly coloured, brilliantly flamboyant species, particularly in Africa. The nymphs of some species aggregate to form hopper-bands in a manner reminiscent of locusts. The majority of species are tropical, notably from Africa and Madagascar. They are poorly represented in the American tropics, where Mexico has the largest number. There are none in the USA or Europe. The **Romaleinae** contains the warningly coloured, slow-moving lubber grasshoppers of the USA and Mexico as well as the giant *Tropidacris* mentioned above. Many species are warningly coloured, rivalling the most spectacular of the African pyrgomorphs in this respect. Most of the 200 or more (probably many more) species are found almost exclusively in the American tropics, although a few beautiful examples penetrate well north into the United States and the Old World has a scattering of species. They occur from deserts to tropical rainforests. Few of them eat grass.

FAMILY EUMASTACIDAE

This is a family of more than 1000 described species of 'monkey-hoppers' which are overwhelmingly tropical in distribution. Many beautiful species with metallic or iridescent colours are found in tropical rainforests. Other kinds are cryptic and mimic leaves or twigs. The head is often noticeably high-profiled, projecting well up above the front of the thorax, and the antennae are usually very much shorter than in the preceding families. The tibiae of the hind legs are often very long, giving eumastacids a gangling appearance. This effect is heightened by their frequent penchant for sitting with their back legs splayed out to their sides, so that the 'knees' are almost (or actually) in contact with the substrate. When in this position the broad insides of the femora often display a conspicuous coloured pattern. There are a number of subfamilies, generally with such an obvious family likeness that their membership of the Eumastacidae can easily be established on sight.

FAMILY PROSCOPIIDAE

A small family containing 130 species of 'stick grasshoppers' in 17 genera. Most of them have a greatly elongated body resembling a brown twig. The

Plate 13 Male grasshoppers are usually very much smaller than the females. This is particularly noticeable in stick grasshoppers of the family Proscopiidae, such as this pair of *Anchotatus* species from dry slopes in the Peruvian Andes. They are not actually mating; the male appears merely to be riding around on the female's back. This family of 130 species is exclusively South American.

Plate 14 This male *Proscopia* species in tropical rainforest in Peru shows graphically the weird shape of the head often seen in stick grasshoppers of the family Proscopiidae.

females are generally very much larger and thicker-bodied than the males, which may look like a completely different species. The long, pointed head is sometimes bizarre in the extreme, with the bulbous eyes mounted high up and the apex topped off by the tiny stub-like antennae, which are the shortest within the Orthoptera. All legs are very slender and the hind pair are not adapted for jumping; nor can any species fly. This family is exclusively South American, although it does just reach the southernmost section of the Panama

Plate 15 The groundhoppers of the family Tetrigidae are some of the smallest members of the Caelifera. The pronotum in many species is greatly expanded to mimic various natural objects. This *Xerophyllum* species from tropical rainforest in Kenya mimics a small dead leaf and lives on the ground.

isthmus. They are found from tropical rainforests to lowland deserts and even in the Andes up to treeless altitudes of 12,000 ft (3600 m) or more, where they live among grass and low vegetation in arid areas of semi-desert.

FAMILY TRIGONOPTERYGIDAE

A small family of rather anomalous grasshoppers from southeast Asia. The genitalia are in the reverse position compared with other members of the

Caelifera. These grasshoppers mimic the foliage of broad-leaved trees, something which is rare in grasshoppers and more typical of the katydids. Little is known about their biology.

FAMILY TETRIGIDAE

A family with 1000 or more members called variously groundhoppers, grouse locusts and pygmy grasshoppers. These are generally among the smallest members of the Caelifera and although widely distributed around the world they are seldom observed in the wild. Many kinds are cryptically coloured to match mossy logs or stony ground. Others bear bizarre ornamentations in which the pronotum is greatly expanded or modified to resemble leaves, stones or spiny twigs. A number of species live beside water and readily dive in when alarmed, swimming well.

SUPERFAMILY TRIDACTYLOIDEA

FAMILY TRIDACTYLIDAE

The 220 species of 'pygmy mole-crickets' are small insects, from less than $\frac{1}{6}$ in (4 mm) to only $\frac{3}{5}$ in (1.5 cm) in length, either all black or variegated in black

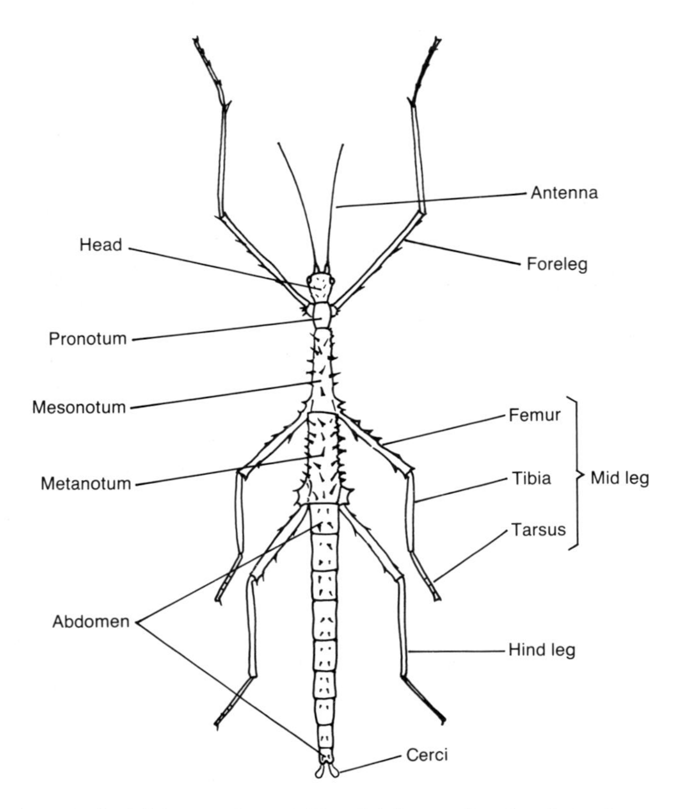

Fig.6 A typical stick insect, the prickly stick insect *Acanthoxyla prasina*.

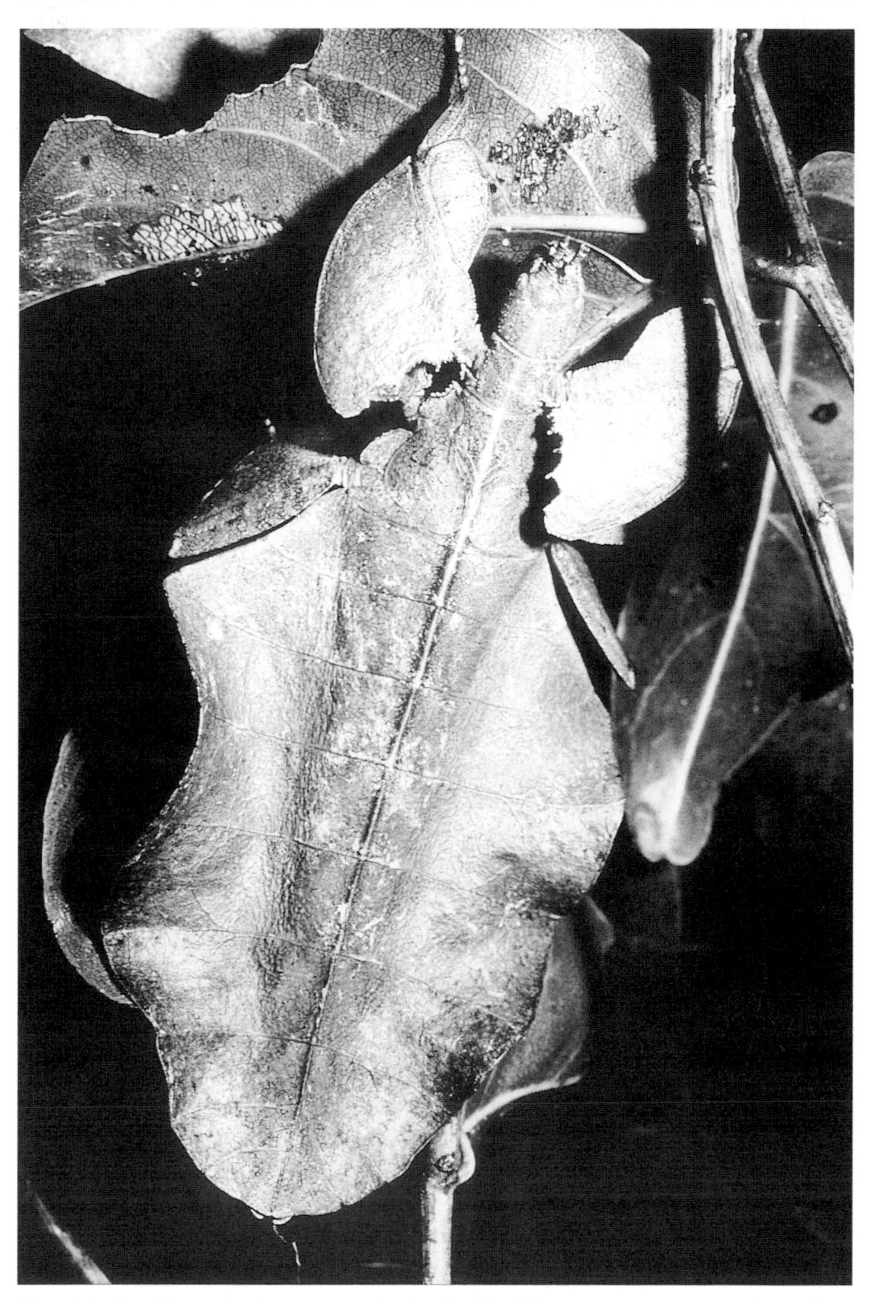

Plate 16 Leaf insects belonging to the family Phyllidae are the only members of the order Phasmatodea which do not resemble sticks. This is a new and undescribed species of *Phyllium* from New Guinea. Like all members of its family it is a superb mimic of a leaf, in this case of the brown dead variety. It has the extremely short antennae typical of a female.

and grey. The rounded head bears well-developed eyes. The tibiae of the front legs are broadened and modified for digging. The hind legs are usually adapted for jumping and the adults are always fully winged. These insects are generally rather gregarious and are found in damp, muddy ground near water; some species swim well. They may dig quite deep burrows for 'nesting' purposes or for hibernation. Being mainly nocturnal their habits have been little studied, but they probably eat small pieces of vegetable matter found in the mud or sand. Both these and the insects in the next family are little known and seldom seen, so this is the sum total of information contained in the present work.

Family Cylindrachetidae

A small family of little-known insects comprising seven species in two genera from Patagonia, Australia and New Guinea. The 'false mole-crickets' are moderately large, stubby-elongate insects with a small head and poorly developed eyes. In some species, the mouthparts are modified to form a stridulatory mechanism. The legs are highly adapted for digging tunnels in sandy soil in damp habitats, from which the insects seldom emerge in daylight.

ORDER PHASMATODEA

This order contains 2500 or so species of insects almost all of which are mimics of twigs and sticks, hence the name walking sticks (used in the USA) or stick insects (used in Britain), members of the family **Phasmidae**. The main exceptions are the leaf insects of the family **Phyllidae** whose flattened bodies resemble leaves. Walking sticks are seldom less than around 1 in (2.5 cm) long and usually much longer, with some species reaching 1 ft (30 cm) or more, making them the longest of all insects. They are mainly shades of brown, but may be black, grey, green (including bright metallic green) or deep metallic blue. The antennae are simple in form and of very variable length, but often noticeably long. The prothorax is short (not elongate as in many superficially similar stick-like mantids) and the legs are all of approximately equal length, usually spindly and not adapted for jumping. Stick insects habitually walk with a strange, hesitant, rocking fore-and-aft action rather like a twig blowing in the breeze. Wings are often absent or greatly reduced but some species do fly, especially the males. The legs are frequently shed as a defensive tactic, but are regenerated (except in adults) at the next moult. Stick insects are widely distributed but most common in the tropics, especially in the Oriental-Australian region. The leaf insects of the family Phyllidae are found from Mauritius and the Seychelles through southeast Asia to northeast Australia and Melanesia. There are only about 20 described species, in which not only the body but also the legs are flattened so that the whole insect resembles a leaf. The antennae of the females are characterisitically very short – even shorter than the rather small head – although those of the males are longer and thinner, and also hairy. In some species the females can stridulate by scraping together lines of teeth on the enlarged third segment of the antennae.

Chapter 2

Courtship and Mating

The prime objective of any insect is to reproduce successfully. This desirable end can only be attained if it is possible to locate a member of the opposite sex of the same species, and then to induce a state of receptivity which will lead to mating and fertilization. There are many different kinds of insects in the world, so any one species will generally find itself immersed in a sea of competing stimuli given off by members of other species. Being wrongly programmed to respond to the incorrect stimulus could be both unproductive and time-wasting in fruitless attempts at copulation with the wrong species. But this rarely happens, for the problems associated with sifting the scrambled input and coming up with the correct answer have been solved in a number of intriguing ways. The devices employed by the orthopteroid insects to avoid confusion and secure a mate of the correct species are many and varied, often being closely linked to the lifestyles of the insects involved.

BLATTODEA

In general male and female cockroaches are initially attracted towards one another by the diffusion of a female sexual pheromone, although in the cockroaches this seems to be of a type with a relatively short range. Once the pheromone has fulfilled its introductory purpose it follows up by triggering a series of responses designed to switch on courtship activities which will eventually lead to copulation.

The courtship sequence in species of the genus *Latiblatella* from Honduras is fairly typical of most cockroaches. When virgin males and females of *Latiblatella angustifrons* meet, and are both in a suitable 'mood', they break the ice with a spot of gentle fencing with their antennae, during which each participant picks up a cocktail of stimuli compounded of the senses of touch, scent and taste. If sufficiently 'turned on' by the female's caresses, the male may respond by rocking his body from side to side. His next step is to adopt a courting posture. He makes an about-turn which leaves the tip of his abdomen pointing towards the female's head. Simultaneously he bows his head and prothorax low and sticks his tegmina (leathery forewings) and wings up on high. By doing this he unveils a special pheromone-producing gland, whose ambrosial products have, until now, been masked by the close-fitting wings. This special gland is situated on a highly modified zone of the male's seventh abdominal segment, very near the tip of his body. The gland consists of a raised lobe decorated with a dense brush of short hairs, bathed by a rather viscous secretion which is assumed to be the pheromone.

A receptive female will be strongly attracted to this male exudation, to the extent that she will stroke it with her facial palps or even actually commence to feed on it, before edging gradually forwards over the male's back, all the while nuzzling him with her mouthparts. By moving into a position with her head

Plate 17 Although most cockroaches initially achieve the mating position with the female atop the male's back she immediately steps off and moves around so that they take up a back-to-back position. This is a species of *Pseudomops*, a small cockroach, pictured mating in daylight in full view on a leaf on the Campo Cerrado near Brasilia in Brazil.

just behind his elevated wings, the female transmits a tactile message to the male, who responds instantly by shoving himself backwards further beneath her. At last he attains his goal – he is in a position from which he can seize the female's genitalia with his own grasping organs. As they finally couple the two insects are, of course, facing in the same direction, with the female atop the male. Shortly afterwards the female steps off the male's back and turns in the opposite direction, so that the two insects now face away from one another, firmly coupled at the rear end by their genitalia. They remain in this position while the male transfers a spermatophore to his mate, a process which averages about an hour for completion.

Females of this and two other species of *Latiblatella* are known to adopt special 'calling' postures, during which the efficiency of diffusion of the male-attracting pheromones is presumably enhanced. Females of *L. angustifrons* assume the calling posture by lowering the front and rear parts of their bodies in a kind of 'hang-dog' attitude, thereby opening up a space between the abdomen and the folded wings. At present the actual source of the pheromone on the female's body is unknown, for there are no obvious modifications such as are found in the male. Such female 'calling' has been noted in three of the five families of cockroaches, and in some species is accompanied by periodic expansion and relaxation of the genitalia. It is thought that these volatile female scents are effective in attracting males over a range of several metres. The main function of the male secretion is to coax the female into a suitable piggy-back position in which the male can grab her and mate.

A number of cockroaches depart strongly from this 'standard' procedure. *Eurycotis floridana* is a 1–1½ in (3–4 cm) long cockroach found in woodlands in the southeastern USA. Neither sex possesses wings (in many cockroach species only the female is wingless, the males being fully winged), so the wing-raising ritual usually seen in male cockroaches is absent. However, in a custom unusual for any kind of insect, it is the female *E. floridana* who actually sets in motion the courtship process. In short, females simply jump on top of any available males, even those which are not displaying the rapid body-vibrations which signify sexual arousal. The stimulus for this almost aggressively sexual female behaviour is probably a pheromone released even by passive males. Nevertheless, actual copulation will not follow unless the male signifies his receptivity to further advances by rapid side-to-side rocking vibrations of his body. This seems to arouse the female to indulge in unusually prolonged feeding on secretions present on the male's first abdominal segment, and leads to copulatory behaviour in both sexes.

Cockroaches of the Madagascan genus *Gromphadorhina* exhibit an even more aberrant form of sexual behaviour, in that production of specialized sounds plays a part, something which is far more typical of many grasshoppers and katydids. Courtship in *G. portensa*, a giant species up to 3 in (8 cm) or more in length, is initiated by the mutual antennal caresses typical of cockroaches. These usually stimulate the male into adopting a specialized posture, with his thorax lifted high off the ground and his abdomen arched upwards. This posture is usually accompanied by the emission of a number of single hisses, produced from a pair of modified spiracles and clearly audible to the human ear. The female usually responds by walking across the male's back, to the accompaniment of thrusting movements by the genitalia of both sexes, and a further burst of much faster hisses which rapidly become trilled, after which copulation often occurs. Males which have been artificially muted fail to achieve copulation, although playing recorded hisses engineers the desired outcome, thus establishing the vital role played by hissing in supplying a complete set of the necessary stimuli to both sexes. It is interesting that the male's custom of luring the female on to the correct position on his back by tempting her with his seductive glandular secretions is absent in these cockroaches. This is because the mating posture is achieved simply and quickly by the male backing into the female, thus dispensing with the preliminary manoeuvring usually necessary in other cockroaches.

Polygamous behaviour probably rules among female cockroaches, but at least one species, the American cockroach *Periplaneta americana*, is monogamous. While most cockroach females do indeed mate a second time, although not until after they have successfully produced their offspring, female American cockroaches just rely on their single inaugural mating to fulfil their reproductive obligations.

MANTODEA

The following statement of 'fact', or something similar, would probably have figured as the obligatory opening paragraph, at any time over the last 20 years, in any account of sexual relationships in mantids.

> Female mantids are notorious for making a meal of their mates. As her suitor perches on her back, devotedly doing his duty, she will usually look round and start casually chewing his head off. This is not the disaster it would at first appear. Far from preventing him from carrying out his vital seminal functions, this severing of the male's head removes inhibitory centres in the brain which actually reduce the vitality of the genitalia; these then perform all the more vigorously as a result.

This nugget of wisdom, almost certainly wrong, was based mainly on observations of mantid sex under captive conditions, during which flaws in the techniques employed by the human observers were seemingly responsible for eliciting abnormally powerful inclinations towards head-pruning behaviour by the participating female mantids. So what is the real situation regarding the state of connubial relations enjoyed by these insects?

Unfortunately, at present, we simply don't really know. Death by cephalic self-sacrifice certainly *is* the occasional fate which befalls mating male mantids, but probably only in instances when something goes wrong to upset the normal tranquil state of coital harmony. In common with most male insects, it is in a male mantid's best interests to sire as many offspring as possible, so rank promiscuity with as many females as possible is a highly desirable objective. As losing his head to his very first partner is not going to give much scope for spreading his genes around, a male mantis is programmed with a set of tactics

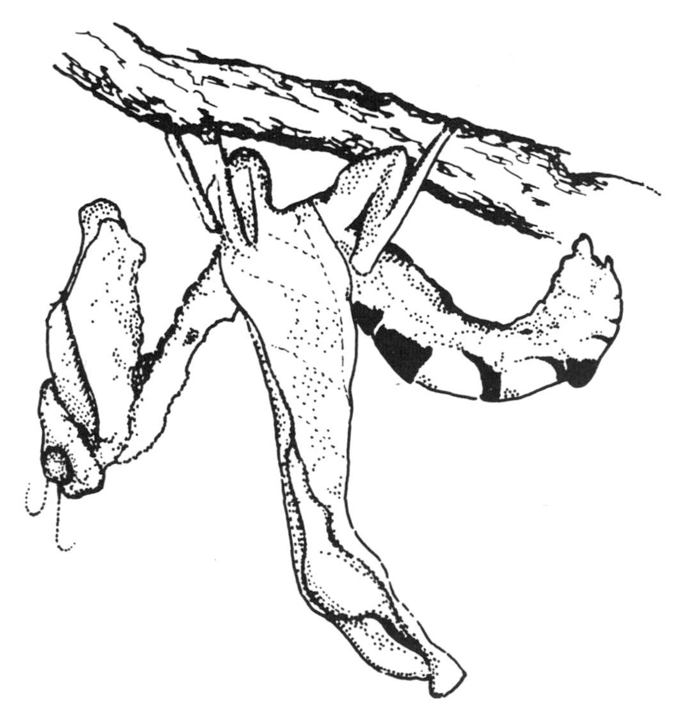

Fig.7 A female dead-leaf mantis *Acanthops falcata* raises her wings and exposes her abdomen to release a sexual pheromone attractive to males.

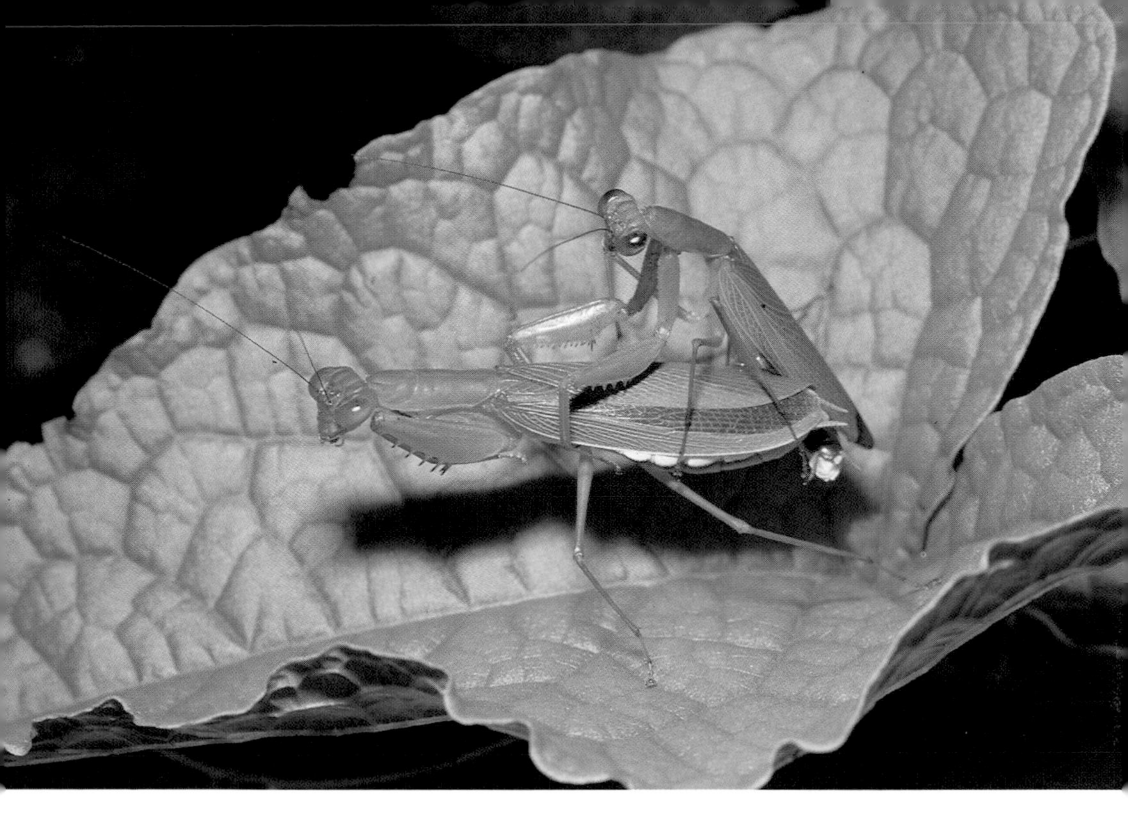

Plate 18 This pair of *Tithrone* species mantids in Trinidad spent 12 or more hours copulating with no apparent break in the connubial harmony. Note how the male's considerably smaller size puts his head a long way back on the female's body, presumably minimizing the risk of tempting her to take a preliminary nibble.

suitably tailored to increase his chances of continued survival during intimate contact with the female. Until he arrives, his prospective mate will have spent her entire life assessing the value of all suitably-sized arrivals as a possible meal. Male mantids are generally much shorter and a great deal slimmer than the females, who may look so different that they seem to belong to another species. This might seem to put the males at increased risk of bullying and eventual decapitation by their burly partners. However, during mating, the head of the dwarf male is placed so far back on the female that 'out of sight, out of mind' may well be an important factor in his continued survival.

However, sexual contacts for many, perhaps most, male mantids could well prove to be established at quite a low personal risk. Take, for example, the actions of the female dead-leaf mantis *Acanthops falcata*, an inhabitant of tropical rainforests in Central and South America. Far from being a reluctant consort, eager to take advantage of her suitor's convenient proximity by taking a few love-bites from his head, the female actually triggers the whole affair off by deliberately diffusing a powerful, species-specific sexual pheromone into the air, a silent yet irresistable siren-call enticing every male in the area to come and visit. And when he does arrive, the spirit of invitation implicit in the message guarantees him a favourable reception.

The wingless *Acanthops* only broadcasts her urgent message for a brief period just before dawn, when the fully winged males, who are rather weak fliers, are at least risk from interception by predators such as birds. Her scent is of such potency that large numbers of males are quickly attracted. The pheromone is released from two shiny black protuberances situated on the upper surface of the female's abdomen, towards its tip. This is fully exposed to the breeze by her adoption of a special calling posture, with her abdomen curved upwards away from her vestigial non-functioning wings.

This transmission serves a number of vital functions. Adult dead-leaf mantids are widely dispersed within the forest and extremely difficult to see, an adaptation designed to prevent their detection by enemies, but equally effective at hindering visual location by prospective mates. The flightless females are obviously incapable of playing an active searching role, and substitute a method which, with its properties of wide dispersal and complete specificity, attract the winged males only of the *desired* species from a considerable distance. Odour-messages have the great advantage of constituting a private communications system which is indecipherable by any but the correct species. In addition, they do not invite the unwelcome attentions of potential enemies such as birds or bats, which may act as 'poachers', homing in on sexual enticements employing more public communications channels, such as sound (see under katydids and crickets below). Finally, when an *Acanthops* male does show up, he encounters a female who has given advance notice of her sexual receptivity by issuing him with an invitation to attend. He therefore has no need to placate a dangerous mate and can simply leap directly on to her back and couple, without resorting to any of the complex preliminaries which may be needed to smooth the path to sexual conquest in some other mantids.

It would be difficult to find two mantids more different in size, appearance and habits than the small, brown, crinkled-looking dead-leaf mantis and the large sleek brownish-green Chinese mantis *Tenodera sinensis*. If he is lucky the male of this species may simply be able to jump straight on to the female's back – but *only* if she doesn't see him coming and he can manage an instant *fait accompli*. But to try this on in full view of the female would be to court disaster. So instead he must proceed more cautiously and prepare the ground by embarking on a specific succession of courtship steps.

When confronted with one another for the first time, the two sexes fix one another with the typical mantis icy gaze. The male then begins an extremely slow and careful 'stalking' of the female which lasts for an hour or two. About half-way through he launches into a complicated sequence of abdomen-waving, which is the curtain-raiser to a sudden flying leap on to the female's back – or into her arms if he miscalculates. Surprisingly enough, this does not seem to be serious, for she usually allows him to make good his blunder without forfeiting his head and to clamber unhindered on to her back. Once in position he bends his abdomen in a sinuous motion which brings it into a kind of stroking, searching contact with the female's genital area, while at the same time beating a tattoo against her antennae with his own. Finally, he twists his abdomen over to one side and the copulatory posture is established. It has been suggested that the courtship-waving of the male's abdomen in a figure-of-eight pattern may generate ultra-sonic signals as his folded wings are rubbed by his abdomen. This 'song' may be perceived by the female's 'ear', which is situated beneath her abdomen. (However, an alternative or perhaps

Plate 19 The arrival of a second male while a pair were already mating probably led to the decapitation of the original male in these large *Polyspilota aeruginosa* mantids in a Kenyan forest.

additional hypothesis speculates that the 'ear' may be used to detect ultrasonic emissions by hunting bats.)

All this effort does not go unheeded by the female. She often responds with pumping movements of her abdomen, she may raise her fully extended forelegs high in the air or even approach the male and stroke his front legs. She thus seems to be an active and positive participant in the whole drama, in which the alluring prospect of recasting her partner's part from mate to meal does not seem to figure. It should also be mentioned that mating can take place successfully without *any* of this complex preamble, but success rates do seem to be much higher if the female becomes fully involved.

The possible consequences of something unexpected interrupting the correct sequence of actions have been observed by the author in Kenya. A female of the large green mantis *Polyspilota aeruginosa* was perched low down among dense grass. She turned her head briefly to gaze up at the author, and then went back to the job which his arrival had interrupted – finishing off the tattered remains of her mate's head. With steady palpitations of her jaws she gradually worked her way down the length of his elongate prothorax, as if she were dealing with a stick of celery. This finished, she then took the first bites from his much chunkier abdomen, methodically whittling away bands of tissue as she chewed away with some purpose. She bent to her task with such effect that within half an hour or so the only remaining vestige of her erstwhile mate was the very tip of his abdomen, where his genitalia were still pulsing away in their final seminal act.

In order to finish off her meal, the femal evinced amazing proficiency as a contortionist, being capable of bending herself almost in half to reach the last succulent morsels. However, one important point has so far been omitted from this account. In order to indulge in browsing so profitably off her mate's substance she was forced to bend *sideways*; not *backwards*, as would be expected if her much shorter mate were perched in his normal position atop her back, with his head set well back from hers. In fact this less risky position was indeed occupied by a male – not her unlucky mate, but a second, intruding male. The steps which had led up to this strange *menage à trois* can only be conjectured, but probably went something like this. Shortly after mating had commenced, a second male had shown up and tried to 'muscle in' on some of the action. Connubial relations had probably been good until then, perhaps after some kind of correctly performed courtship procedure by the original suitor, producing a trance-like state of acceptance in the female. The sudden arrival of a second male, probably followed by an attempt to depose the original incumbent, seemingly ended up with the first male pushed over to one side. With her quiescent state now thoroughly disturbed, the female was presented with a head perched invitingly close to one side of her own, encouraging an exploratory nibble. The final result was plain to see.

The arrival of the second male, who had detected the pair among dense grass, suggests the dispersion of a female invitation-pheromone, probably implying risk-free mating for the first-comer. That interrupted matings do not necessarily lead to head-amputations was noted by the author's observations of a similar event in Trinidad. This revealed just how incredibly indifferent a female mantid can be to even violent rockings of the sexual boat, and exposed a degree of placidity in the female nature very different from the 'conventional' view quoted at the start of this section.

The drama involved three individuals of a pretty little green and white *Acontista*, whose rather plump females often perched on white flowers on the edge of the rainforest. The males are quite different, being lightweight, slim-bodied, fully winged dwarfs. One of these sylph-like suitors was ensconced atop his corpulent and quiescent mate when a second male landed on the leaf nearby. This in itself posed an interesting question. Had he glimpsed the mating pair? Or had be been lured to the spot by a female pheromone? Or perhaps a combination of the two? Whatever the answer, he quickly buckled down to the job of unblocking his access to the female by getting rid of the male-in-residence.

One important point was immediately obvious. The intruding male was well aware of the implications of the presence of the first male, not attempting to sit alongside him and make fruitless efforts to couple with the female's fully occupied genitalia. This unrewarding ploy is quite common in other groups of insects in which 'troilism' is seen. Instead, the male *Acontista* scrambled up over the female's head to face his opponent. What followed was a kind of slow-motion fight in which the intruding male strained to depose his rival by shoving him backwards off the female. This struggle for supremacy was notably lacking in the viciousness observed by the author in similar circumstances in longhorn beetles. On the other hand, the reaction of the female mantis was anything but violent. During counter-thrusts by the resident male, his opponent's rear end was several times pushed back across the female's face; yet this failed to elicit any desires to take a quick bite. In the end, the original male managed to retain his seat and the intruder flew off. Both males came out of it completely unscathed, despite the violent interruption of the normal mating sequence, which could theoretically have put both of them at risk.

Summing up, it seems probable that although in some (perhaps all) species, male mantids are programmed with a full repertoire of courtship moves, these are probably not often required in nature. This conclusion is based partly on the author's own observations of troilism in the wild (not all reported here), which provide persuasive circumstantial evidence for the frequent use of invitation-pheromones by soliciting females, who are thus willing partners, not requiring any complex preliminaries to get them 'in the mood' for mating; and partly on the probability that under natural conditions the far more active and sprightly male may often be able to dispense with formality and simply take a flying leap on to the unsuspecting female's back, relying on his much shorter body to keep his head intact once he's in position.

ORTHOPTERA

TETTIGONIIDAE

Mating in katydids is difficult to observe, for the action usually takes place under cover of darkness, and despite seeing many thousands of katydids around the world the author has only once come across a pair copulating. Kaytdids employ sound rather than scent as a means of long distance communication between the sexes. The males are the songsters, although the females also sometimes reply with a far softer sound. The short, brisk, rather subdued chirps of large numbers of male dark bush crickets *Pholidoptera griseoaptera*, uttered from deep in a bramble bed, are a familiar sound as dusk falls near the author's home in central England. The sound is produced by drawing a 'scraper' on one wing against a 'file' on the other. In the tropics the male songs are often amazingly penetrating and strident, usually being delivered from a prominent stage on top of a broad rainforest leaf. Tracking down these songsters can be frustratingly difficult, for they are master ventriloquists and at any one moment the song can appear to emanate from positions which are yards apart. Yet finally just a single singing male is tracked down, perhaps quite far removed from where the searcher had been convinced he was sitting.

The acoustic displays of the American meadow katydid *Orchelimum gladiator* are perhaps characteristic of species which lurk deep among dense vegetation. Aggregations of males sing virtually throughout the daylight hours in a continuous serenade reminiscent of the courtship leks formed by certain birds and frogs. The individual song lasts only a few seconds and consists of a few simple ticks alternating with a buzz, but song follows song with such rapidity that a non-stop chorus is formed. These songs act as beacons calling in the non-singing females, as well as establishing a fixed distance, usually of not less than $1\frac{1}{2}$ yd (1.5 m), between singing males. However, a calling *Orchelimum* will frequently desert his song-post to track down his nearest singing rival and engage him in a no-holds-barred wrestling match, during which the antagonists indulge in a great deal of turbulent biting and kicking. Such overtly aggressive behaviour is so far only known in males of this genus, but the phenomenon of males zeroing in on one another's songs seems to be common within the family.

Most male katydids are able to 'sing' loudly in this way, but there are exceptions. Male European oak bush crickets *Meconema thalassinum* can generate only the most muted of sounds with their wings. They make up for this by employing a kind of tap-dancing method which is probably uniquely theirs. The wings are elevated almost to the vertical while one of the hind feet is drummed rapidly in short bursts agains the leaf. Meanwhile the spare hind leg is stretched out backwards to act as a brace. The volume of sound thus produced is audible to the human ear from a range of several yards, while probably further by the female bush cricket.

Males of some tropical katydids produce sound in two completely different ways. This acoustic dichotomy has probably evolved as a counter-measure in the on-going war between bats and their prey. It is well known that bats home in on the mating calls of certain tropical forest frogs, and it seems likely that they do the same with serenading katydids, for up to 40% of the detritus in roosts of one species of bat was found to consist of katydid remains. Any strategy which will reduce the risks of ending up inside a bat-dropping without compromising too far the ability to attract a female is obviously worth employing.

The calling repertoire of the male *Copiphora rhinoceros* thus incorporates two parts, normal stridulation and a broadcasting system known as tremulation. The males usually give their bipartite performance from the underside of a large robust leaf such as that of a *Philodendron*. A spell of stridulated buzzing is followed by a period of silence, which is then succeeded by a session of tremulation. This special method of restricting self-advertisement to the attention of females only, thereby reducing the risk of being picked off by a marauding bat, involves the use of percussive technology. Raising his body higher than normal off the leaf, the male rapidly vibrates it from side to side in a stereotyped display. This imparts a high frequency vibration both to the leaf and to the connected surrounding vegetation, transmitting a message which is accessible to females settled on nearby vegetation but unavailable to flying predators. By alternating between stridulation and tremulation, the males reduce the chances of undesirables tapping into their communication channels and minimize the risk of detection by anything other than their own females.

Once a female has shown up, the male *C. rhinoceros* dispenses altogether with the high-risk stridulations and relies entirely on tremulation for success in courtship. As with other bush crickets, the larger female mounts her smaller mate and copulation proceeds with the male hanging beneath the female, often supported entirely by their joined genitalia. This is in contrast to grasshoppers in which the male (who is also smaller) rides on his mate's back.

The mating process in katydids is often prolonged, perhaps lasting for several hours, for a very good reason. Towards the end of copulation – during the last five minutes of a nearly four-hour mating period in *C. rhinoceros* – contractions of the male's abdominal muscles eject a large white spermatophore which becomes firmly attached to the underside of the female's abdomen, near the base of her prominent ovipositor. The production of the spermatophore

Plate 20 During copulation in katydids, the smaller male is normally carried beneath the body of the larger female. Note the long upwardly-pointed ovipositor of this female *Dichopetala* species, mating in desert in northern Mexico.

Plate 21 This male 'tizi' *Ephippiger* species katydid from France is in the act of calling to attract a mate. He has raised his vestigial forewings and is rapidly vibrating one against the other to produce a buzzing sound. Males of this genus also 'tremulate', vibrating the whole body in order to produce sympathetic vibrations on the plant upon which they are perched.

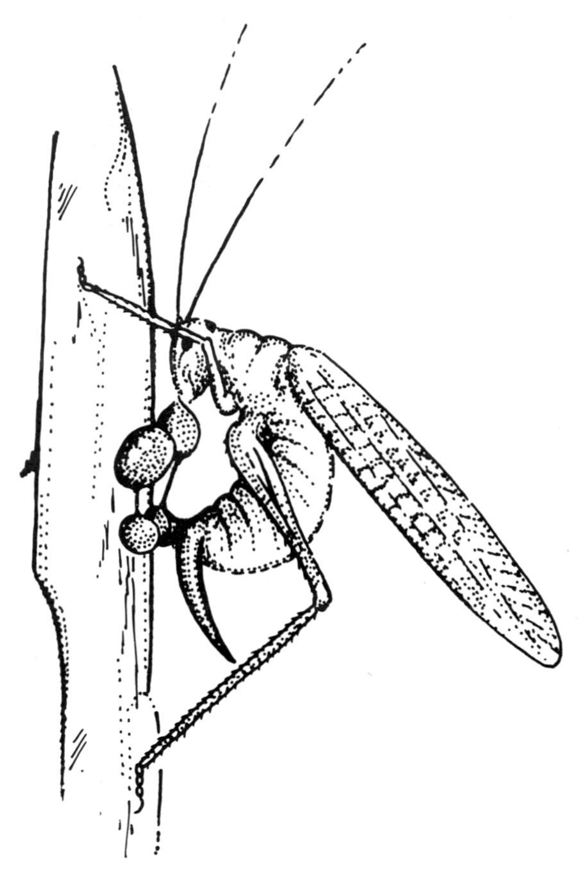

Fig.8 A female katydid *Bliastes insularis* eating the spermatophylax of the spermatophore which has just been transmitted to her. (The middle nearside leg and the right-hand legs have been omitted from the drawing for clarity.)

usually represents a substantial investment in resources by the male, for it can be amazingly large – $\frac{4}{5}$ in (2 cm) long by $\frac{2}{5}$ in (1 cm) wide in *C. rhinoceros* and up to 30% or more of body-weight in males of the European *Ephippiger bitterensis*. The spermatophore consists of two parts – a sperm-package which is inserted into the female's genital cavity, and a much larger, white, jelly-like blob called the spermatophylax, which remains conspicuously attached to the outside of the female.

Virtually the first act of the female after a mating pair has separated is to arch her body so as to bring her mouthparts and abdomen-tip into conjunction. She can now eat her 'wedding present' – the spermatophore. The extra bulk of the spermatophylax has been added for just this purpose, being extremely rich in nutrients which eventually find their way into the female's ovaries and thence to her mature eggs. It has been found in the Australian katydid *Requena verticalis* that the larger her post-coital meal, the greater the quantity and size of the eggs produced by the female. However, not all male katydids make such a generous investment in their offspring's prosperity, for some donate a spermatophore representing only as little as 2% of body-weight.

GRYLLIDAE

Male crickets are accomplished songsters, often having a varied repertoire of calls to suit the particular circumstance. Although a familiar sound in warmer parts of the world, the chirping of a cricket is rarely heard in the British Isles, for only one species, the wood cricket *Nemobius sylvestris*, is reasonably common and its song is subdued and seldom noticed unless one is listening specially for it.

Yet despite its lack of vocal authority the courtship procedures in the wood cricket are particularly interesting, being far more complex than is normal within the family. Males sing mainly during the day, and the individual calling of many males usually joins together to form a continuous chorus. When a male spots a female attracted to his serenading he turns the volume down a trifle and begins a series of jerking to-and-fro movements. This results fairly quickly in the appearance of a spermatophore at the tip of his abdomen. He then steps up his jerking rate until the female is sufficiently aroused to crawl on to his back, upon which he falls silent and attaches the spermatophore near the base of her ovipositor. The pair now quickly split up and it would seem that

Plate 22 This female armoured ground cricket *Acanthoplus armativentris* (Tettigoniidae) has recently mated and is now carrying the large spermatophore donated by the male. This species is very common in dry areas of southern Africa.

Plate 23 This male *Gymnogryllus elegans* produces a piercing song of painful intensity by amplifying his call using an earthen wall at the entrance to his burrow. He only sings at night and stands with his rear end facing outwards through a gap in the 'amplifier'. This species is very common in certain tropical rainforests in Malaysia.

everything has been satisfactorily concluded. But then the male resumes his jerking actions and the female again responds by crawling on to his back. Now she is in far less of a hurry to dismount and lingers there for some time – perhaps several minutes – busily absorbed in licking a small area on the male's right forewing. The object of her attentions is a patch of tiny hairs coated with a secretion which the female finds very attractive. She has to take care in her delicate nibblings, however, for if she gets carried away and takes too enthusiastic a bite, the male will promptly throw her off.

With her feeding completed, the female again dismounts and remains nearby, often occupying herself by rubbing off the spermatophore she received earlier and eating it. After a few minutes, the male produces another spermatophore, which is notable for being more than twice as large as the first one. Ignoring the female completely, the male wanders around for a while, carrying the spermatophore, until he suddenly strikes up a song once more and initiates another set of jerking actions. They mate and the female receives this second spermatophore, whose eventual fate is also to be rubbed off and eaten, but not until it has performed its vital role of fertilization.

The significance of this lengthy and quite complex procedure is hard to assess, for other related crickets manage perfectly well with far fewer formalities. In the American *Hygronemobius alleni*, a male confronted with a female produces a short burst of song. He then positions himself behind her and lashes her body vigorously with his long, hair-like antennae. After several minutes of this, a spermatophore appears and the male encourages the female to mount and take it by pushing forward and softly chirping, all the while keeping up his antennal lashings. The act of mating is brief and the female again finishes off the whole sequence by eating the spermatophore, a practice which seems to be standard in the crickets.

Male crickets are often strongly territorial and may sing from a burrow specially constructed for the purpose. It seems that stentorian abilities are a positive advantage for a male cricket, for the more piercing his call, the more likely he is to attract a female. Certain tropical crickets go to the lengths of building special structures to amplify their output. The distinctive black and white male *Gymnogryllus elegans* constructs a little mud wall around the mouth of his burrow, leaving a gap in one side where he can stand and call. These little earthen megaphones are a common sight dotted about the forest floor in parts of peninsular Malaysia, but their function would be a complete puzzle were they only observed during the day. At night, however, the mystery is solved in no uncertain manner, for as dusk falls the male sits with his rear end facing the gap in the wall and produces a song of such ear-splitting pitch and volume that getting close enough to take a photograph is an unpleasant experience without the comfort of ear-plugs.

The American field cricket *Gryllus integer* is common on suburban lawns and pastures and has been extensively studied. Males sing at peak volume, both to defend a territory against intruding males and to attract females. However, dominant males, which are usually the largest ones with the noisiest songs, will actively seek out and assault males who are preoccupied with the mating process. The bellicose behaviour of these 'sexual muggers' is justified by the likelihood that it will lead to more frequent sexual conquests and the passing on of more genes. However, there are also considerable risks involved in being a macho male field cricket. Although his lusty song certainly brings females at a run, it also attracts stealthy 'satellite' males, who silently lurk by the caller's burrow in an attempt to waylay females attracted by the tenor quality of singing. As it happens, these taciturn would-be Lotharios seldom succeed in their undercover sexual ventures. But neither do they succeed in attracting a far less welcome visitor who, like the females, is also strongly attracted to a mega-voiced male. This is the small yellow red-eyed fly *Euphasiopteryx ochracea*, whose females utilize the homing-beacon so helpfully provided by the singing males to simplify their job of locating a host. These flies contain a batch of larvae, all primed and ready to go as soon as their mother deposits them on to a male cricket. Their first chore is to bore into the cricket, whose internal tissues are then eaten away over the next week to satisfy the voracious appetites of his uninvited guests. He shows few external signs of his inner problems, but within a short time of the fully grown fly larvae bursting back out into the outside world to pupate, their former host withers and dies. Even if the female fly cannot manage to drop her larvae directly on to a cricket, he may still be doomed, for she deposits them in the grass near his burrow and they attach themselves to his body as he goes by. Despite the increased probability of coming to this sticky end, it still seems to be preferable for a male field cricket to call as loudly as possible and attract a mate, rather than slink silently around as a satellite, to remain a secure but impotent virgin.

The slender, delicately built males of the tree crickets (Oecanthinae) also sing, usually from a prominent position on a leaf. Being rather small and weak, these insects cannot produce the kind of volume typical of the larger crickets. At least three South African tree crickets, including *Oecanthus burmeisteri*, solve this problem by adapting part of the environment to improve matters by becoming a natural amplifier. The male bites a special pear-shaped hole in a leaf, suitably tailored so that his elytra, or wing-cases, are a snug fit when he occupies the hole in his calling position. It has been shown that by exploiting the natural resonating qualities of the leaf in this way he can increase the volume of his song by a factor of three or more. Male tree crickets are generally strongly territorial, and if another male bids to trespass, a territory-owner will challenge the interloper by switching from his normal song to a special aggressive song. If the bout of vocal fencing which then ensues does not settle the matter of ownership, the adversaries will usually quickly resort to violence, lashing one another with their long antennae, biting, kicking and struggling until a clear victor emerges. Such tussels over squatters' rights soon lead to a definite dominance hierarchy among the adult males and also ensure that they are distributed throughout the habitat, making liaisons with the available females less dependent on mere chance.

GRYLLOTALPIDAE

A few years ago the author spent 20 minutes on a humid tropical night trying to track down the source of a continuous low buzzing coming from a lawn. It proved surprisingly difficult to decide exactly *where* the sound was coming from, and success was only eventually assured by applying an ear to the ground and, in this awkward posture, sweeping back and forth across the grass. Such is the incredible ventriloquial quality inherent in the call of the male mole-cricket, although presumably only we humans suffer such problems in pinpointing the caller – female mole-crickets probably head straight in with unswerving accuracy.

Britain's single species of mole-cricket is virtually extinct, so the song is seldom heard there, but these are common enough insects in warmer countries and their continuous churring buzz is as much a component of the tropical night as the raucous croaking of frogs and the chirping of crickets. Male mole-crickets park themselves at the entrance of their burrows while intoning their rather monotonous songs. They always sit in more or less the same spot each time, facing inwards towards the burrow. This habit is neither whim nor accident, but for a good reason. Mole-cricket burrows are not merely randomly shaped holes in the ground but sophisticated sound-systems, exploiting the laws of physics to magnify the singer's efforts as efficiently as possible.

The calling-burrow of the European species *Gryllotalpa vinae* takes the form of a Y-shaped double-horn. The internal surfaces are rendered to a fine degree of finish, which is probably achieved by some form of smoothing with the rounded contours of the head and thorax. When singing, the cricket positions itself with its wings just in front of the specially constructed bulb at the lower end of the burrow. By calling from this position, the extra volume of the bulb functions as a filter which suppresses sound-losses into the rest of the burrow leaving more sound to pulse forth into the night air from the twin horns.

The singing of certain males seems far more attractive to females than the output from competitors. It seems likely that the choice of burrow-site may be important, for burrows located in moist soil emit more volume than those in drier sites. Females are attracted to moist soil in which to lay their eggs after mating, so males who have located themselves in damper areas enjoy heightened success in the mating game. The actual courtship process in mole-crickets is fairly simple. With a female nearby the singing male rocks from side to side, his wings dropped down by his sides. It may require up to half an hour of this performance before the female is minded to climb on to his back, when a spermatophore is attached to the tip of her abdomen.

STENOPELMATIDAE

In this family males often advertise themselves by drumming with their bodies against the substrate – against branches and leaves in *Gammarotettix* or against the sides of an earthen burrow in species of *Stenopalmatus*, such as the American Jerusalem crickets *S. intermedius* and *S. nigrocapitatus*. In the latter species, the opening moves in courtship seem to be surprisingly violent, for the two sexes appear to attack one another in deadly combat. With mandibles agape they engage in several brief but vigorous wrestling matches before attaining the mating position, with the male atop the female and facing towards her rear. He grips her hind legs in his mandibles, curves his abdomen over sideways to connect with her genitalia and instantly passes across a large spermatophore accompanied by a quantity of fluid, a process which seems to evacuate a surprisingly large proportion of his abdominal cavity. The whole affair only occupies five minutes and the female quickly begins to eat the bulky spermatophore. Immediately after mating, males of *S. nigrocapitatus* run a considerable risk of being assaulted by their former mates and utilized as a post-coital feast. The females' cannibalistic temperament may account for the shortage of adult males, which has been noticed both in the field and in collections.

Some stenopalmatid suitors seem to enjoy much more amicable contacts with the opposite sex. The large males of the New Zealand tree weta *Hemideina crassicruris* may accumulate harems of females around them in their refuge-galleries in holes and cracks in trees. However, relations with other males are noted for being argumentative and characterized by frequent attempts to take over a refuge by evicting its occupant, usually another male but occasionally a female. This is accomplished by dragging out the rightful incumbent by one of its rear legs. Once outside, the ultimate winner between two males is usually decided by a head-on clash accompanied by gaping of their massive jaws. Just a single head-butt seems to be an adequate decider in these contests, persuading the losing male to turn tail, leaving his former home, complete with any resident females, to be occupied by the victor.

ACRIDIDAE

In contrast to the situation in many of the above insects, sophisticated courtship in grasshoppers is often absent and a male may be able to secure sexual success by no more complicated an expedient than simply jumping on top of a receptive female. Grasshoppers often occur in dense aggregations of males and females, where their familiar chirping stridulations serve both to space out the

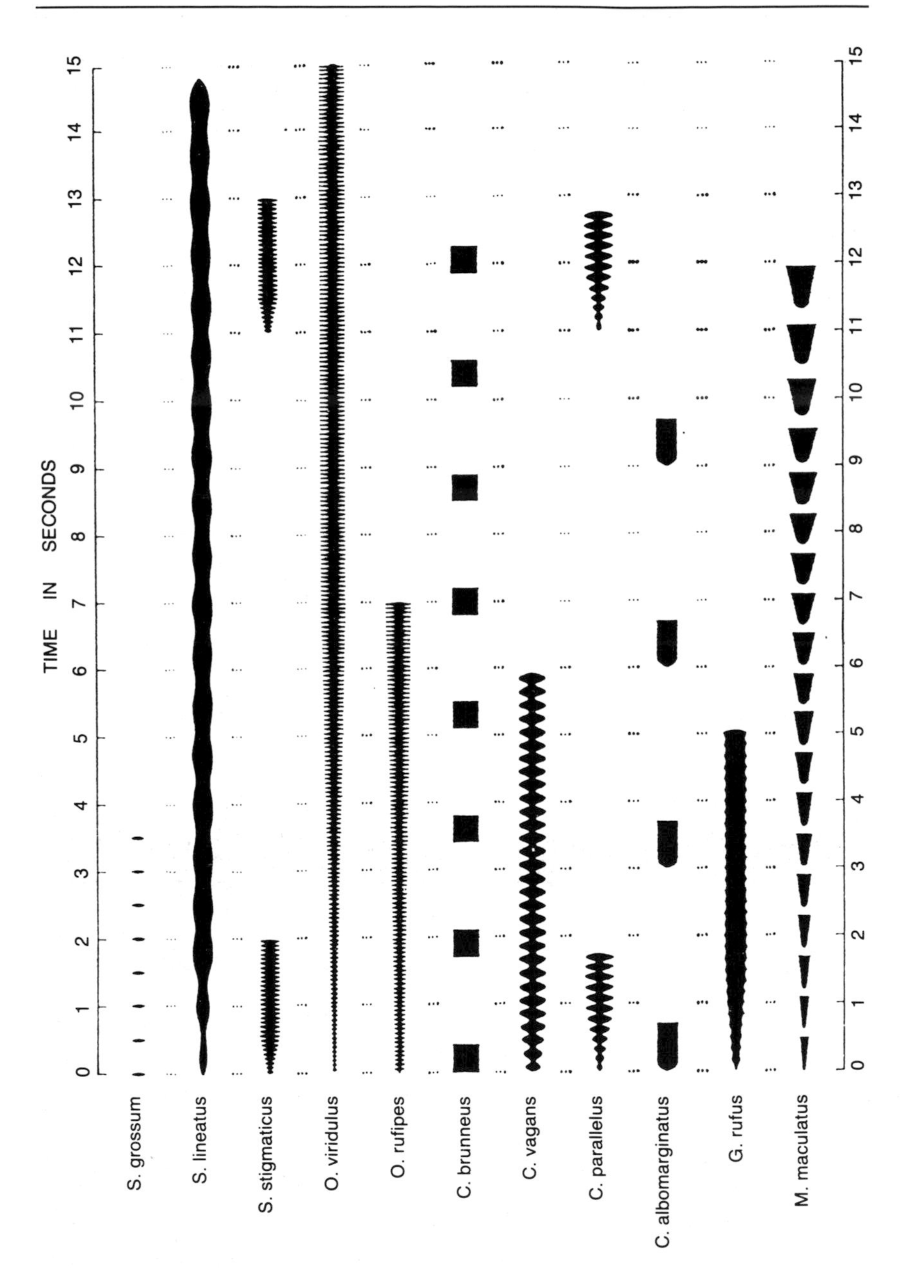

Fig.9 Diagrams of the songs of the British grasshoppers as compiled by Dr David Ragge of the Natural History Museum, London.

males and to attract females. The song serves as a species-identifier, so it is distinct for each species and with practice it is possible to pick out the different grasshoppers present in an area on characters of song alone. Males tend to respond to the songs of nearby males, so a chorus of chirps may be followed by a period of silence, before another male strikes up and starts off the whole sequence again. Many females also stridulate, but more softly than the males, although even this may be sufficient to attract some masculine company. In consequence, the sexes are seldom presented with any difficulty in locating one another, this being further facilitated by the good eyesight and diurnal habits typical of the group. This leads to colour playing an important role in species-recognition and courtship in many grasshopper species. Yet, this said, when a male and female do meet, it may be the circumstances which decide between male success or failure, rather than his courtship efforts, even where these constitute the most complex so far recognized.

Males of the American grasshopper *Syrbula admirabilis* are prone to suffer from this paradox. They have at their service the record number of 18 distinct movements during courtship. These involve intricate co-ordinated actions of the legs, palps and wings in a complicated and lengthy routine. And yet this rich portfolio of seductive virtuosity is all totally superfluous when the male is confronted with a female who has been attracted by his chirping. All he needs to do then is jump on her back and mate. And when he *does* launch into his

Plate 24 This male *Gomphocerus (Aeropus) sibiricus* (on the right) is engaged in a series of body- and leg-movements designed to seduce the female on the left. What he does not do, however, is bring into play his front legs whose tibiae are greatly expanded and flattened to form two 'flags'. It would be reasonable to assume that these have evolved to serve as courtship signals, yet there is currently no firm indication that this is so. Photographed in the Swiss Alps.

Plate 25 Male grasshoppers are generally smaller than the females, sometimes very much smaller, as instanced by this mating pair of *Prionolopha serrata* in a Peruvian rainforest at Tingo Maria. The female is a very large grasshopper with well-armed back legs. Male grasshoppers normally perch on top of the female's back while mating. This is not possible in this species in view of the particularly great disparity in size and the female's deep cross-section at her rear end.

special display, it seems condemned to a cool reception, for the female inevitably seems to ignore all his posturing and remains steadfastly aloof. One can only assume that in this species the courtship performance has evolved as a useful but seldom effective stand-by strategy in case calling fails to attract a receptive mate.

The European mottled grasshopper *Myrmeleotettix maculatus* also rates as one of the top courtship performers, although not perhaps in the same league as the previous species, having only ten different body movements. When confronted by a female, the male greatly speeds up his chirping rate, accompanied by gentle swaying of his body and sometimes punctuated by slight 'ticking' sounds. The complexity of the procedure begins to escalate as he jerks his hind legs and flicks his antennae backwards. He now changes vocal gear and produces a modified version of his song, characterized by each chirp being slightly longer than before. These new chirps are split into two elements, the first part of which sees vibration of the rear legs in a higher position and the second part in a lower position. The choreography is backed up by continuous side-to-side swaying of the body. Finally the rear femora shift into a steady vibration which produces a continuous, sometimes pulsed sound, synchronized with swaying movements of the head and the extension and flexing of the rear legs. Mating is then attempted as the male hops on to the female. If she is unwilling, she kicks him off bodily with her powerful back legs, a positive and unmistakable gesture of rejection which seems to be pretty much the standard method among female grasshoppers in general.

The critical role played by song in keeping species separate can be seen in

Plate 26 The normal grasshopper mating posture is obvious from this photograph of *Eupropacris ornata* in Kenya, with the smaller male sitting on the back of the larger female. In many cryptic grasshoppers, the males and females may be slightly different in colour and pattern, corresponding to the general variation within these types. The species illustrated is warningly coloured and both sexes bear an identical uniform. This is because it is in the interests of all chemically protected, warningly coloured insects to look as similar as possible, thus requiring enemies to memorize fewer 'keep off' patterns.

the European grasshoppers *Chorthippus brunneus* and *C. biguttulus*. Externally the two species are very similar, as is the courtship behaviour, and in captivity the two species hybridize freely to produce healthy offspring. However, in the wild, such crosses appear to be extremely rare, even though both species often occupy the same habitat. The key to this separation seems to lie in their different songs, which provide far more of a stimulus to mating in the 'correct' species than in the 'incorrect' one. This opinion is backed up by experiments which have shown that a female grasshopper can often be fooled into mating with a male of the wrong species who has been artificially prevented from stridulating, and then rendered acceptable by substituting the taped song of a male belonging to the female's own species.

Despite the importance of song in the sex-lives of many grasshoppers and the strong association between grasshoppers and song in the human mind, there are many species, perhaps the majority, in which the males manage perfectly well without uttering a single note, relying on sight and scent to locate the females. Males of the rather drab American species *Melanoplus tequestae* seem to have dispensed with courtship finesse entirely and seek to lay claim to the female's acquiescence by dint of sheer physical staying-power. An ardent male simply jumps straight on top of any likely female and then, using the same principle of winning the prize as a cowboy riding a bucking bronco, attempts to stay the course by hanging on tightly while the cavorting female performs a couple of back-flips and generally does her utmost to pitch him off. If he manages to stay seated after from three to six leaps, then he's won the day, for the female's will to resist seems to fade rapidly and she spreads her rear femora slightly so that he can couple. Once he is firmly in the saddle, mating can last for any time between a few hours to a day and a half, although the latter period may be rare in the wild where disturbance by other grasshoppers is likely. The mating period in *Melanoplus sanguipes* averages 45 minutes, which is probably near the norm for most grasshoppers. Males of this species can

Plate 27 Many grasshoppers have brightly coloured areas on the insides of the back legs. These can be exposed to warn off rivals or attract a mate. This is *Xanthippus corallipes* from Arizona.

readily distinguish between the various acceptability-ratings of females i.e. whether they are immatures, virgins or already mated. It seems likely that in this and other grasshoppers it is a female pheromone which is involved in conveying information concerning her marriageable status. During the actual process of copulation the male passes across to the female a series of very small spermatophores. In the course of its four-hour copulation, a male desert locust *Schistocerca gregaria* produces between six and fourteen spermatophores, each about $\frac{1}{16}$ in (1.5 mm) across.

Many species employ a far more visual courtship and sport conspicuously coloured areas on the insides of their hind legs. These 'sexual logos' are normally concealed, but they can be displayed during courtship or in competitive encounters between rival males. In the subfamily Oedipodinae, solitary males advertise themselves via an aerial display involving either a loud cracking of the hind wings or a colourful display of these wings, or a combination of the two. The hind wings are very decorative, usually shades of red, pale blue or yellow and also serve a (possibly primary) defensive function. The cracking sound is called crepitation and is different in each species.

Vision plays the major role in mate-location and identification in numerous species of brightly coloured metallic grasshoppers which inhabit tropical rainforests. In Costa Rica there is a whole group of such species, including the green-and-gold solanum grasshopper *Drymophilacris bimaculata*. The $\frac{3}{4}$ in (2 cm) long wingless adult is a brilliant green complemented by metallic gold embellishments of the head, prothorax and genital area. The long black antennae are eye-catchingly tipped with white, while the male's rump is decorated with two prominent 'rear lights' in the form of two golden spots. Males initially locate mates by searching the narrow range of food-plants (Solanaceae only) and announce their presence by drumming on the leaf with their back legs. The female drums in response and from then on the pair stays together, using solely visual methods, presumably locked into their bright species-specific regalia. In some grasshoppers there is a marked difference between the colours of the two sexes, often equally as striking as the disparity between the gaudy male and dowdy female plumage of some birds. For example, in the New Guinea grasshopper *Sphaerocranae bipartita* the males are blatantly conspicuous objects memorably attired in a striking uniform of black and yellow bands; the females are green!

PHASMATODEA

After the refinements seen in many of the above insects, it may come as an anticlimax to learn that courtship in stick and leaf insects is more or less absent. There is a particularly marked difference in size and weight between the two sexes, with the males being shorter, lighter and generally noticeably more puny than the females. Males almost always have well-developed wings and fly to their sexual rendezvous, probably navigating towards pheromones released by the females. Once he has located a female, the male simply climbs on to her back and twists his abdomen down and around one side of hers in order to contact her genitalia. In this position he is carried around by the female and may accompany her in her feeding activities.

Plate 28 These charming little grasshoppers mating in a Mexican rainforest illustrate a number of interesting points. The male is obviously smaller than his mate and is in the 'classic' mating position. He exhibits an extreme degree of sexual dichromatism, being far more brightly coloured than the rather drab female, something which is more familiar in many birds and relatively rare in grasshoppers. They not only belong to a new and undescribed species, but also probably to a new genus. This is not surprising, as there are undoubtedly very many species of grasshoppers and other orthopteroids waiting to be discovered and described. The author has already found several unknown species just by spotting them sitting around on leaves.

The amount of time which the pair spends together is incredibly variable. *Cyphocrania gigas* mates for 12–14 hours, while males of such species as *Timema californica* and *Anisomorpha buprestoides* may spend the whole of their adult lives riding atop their mates, although actual copulation only takes place at intervals. The all-time record is probably held by the Indian stick insect *Necroscia sparaxes* which may stay coupled for as long as 11 weeks. The males of some species are rather ephemeral and soon die in harness, leaving the field clear for other males to replace them. Transfer of semen can either take place directly or via a spermatophore. Species which habitually indulge in repeated matings, such as *Extatosoma tiaratum*, exercise sperm competition, whereby nearly 98% of the sperm left behind inside the female's genitalia by a previous male will be

Plate 29 This small grasshopper *Sphaerocranae bipartita* from New Guinea displays even more extreme differences in colour between the male (on the right) and the female (on the left). The male appears to be warningly coloured while the green female is cryptic.

displaced by the sperm of a subsequent male. In order to give his own sperm at least a reasonable chance of securing a place at the head of the queue, the male *E. tiaratum* makes a virtue of promiscuity, devoting his life to inseminating as many different females as possible, as well as making absolutely sure by mating several times with individual partners. Avoiding sperm competition is probably the reason for the lengthy periods spent together by some of the above species. By sticking with his mate, the male is in the ideal position to repel boarding-attempts by any rivals, until he is assured that his and only his sperms have fertilized the female's eggs.

Gate-crashing males intent on mate-theft usually show little sign of determined endeavour and make a few futile attempts to engage their genitalia before giving up and leaving the original male to get on with it. However, the males of certain species seem remarkably reluctant to admit defeat – or perhaps females are sometimes thin on the ground – for in the Seychelles Islands a female of *Carausius alluaudi* was once found with no fewer than six suitors hanging on to various parts of her abdomen. It is probably normal for resident

Plate 30 Male stick insects are generally very much smaller than their females, such that they may even seem to belong to a completely different species. This is very evident in these mating *Ctenomorphodes tessulatus* in eucalypt woodland in Queensland, Australia.

males to mount a passive defence of their mates, relying on their own occupation of the female's genitalia to thwart further engagements by later arrivals. This laid-back attitude is not typical of males of the American stick insects *Diapheromera veliei* and *D. covilleae*, which are far more forceful in defence of their property rights. At the mere approach of a possible competitor a resident male seeks to neutralize the threat by using his clasping organ to bend the tip of his mate's abdomen down upon itself. It thereby ends up in a position such that the normal genital attachment-site is inaccessible to the rival male. If this tactic fails, and the intruder presses home his attentions and attempts to copulate, the dispute may escalate into actual physical combat.

The pair embroiled in the dispute dangle from the tip of the female's abdomen, supported only by their clasping organs. Each then proceeds to rain blows upon his adversary, using wide sweeps of his front legs. The contest may last anything from a few seconds to several minutes before one of the participants lets go and the winner claims the female. Resident males seem to have a poor record in seeing off the opposition, for in 70% of observed encounters resulting in combat, the challenging male emerged as victor. Males of both these species have a special, well-developed, hooked 'fighting spine' situated on the mid femora. This armament is tough enough for males of *D. covilleae* to be able to use it to draw blood from an opponent. The clasping organs at the tip of the male's body are also particularly well-developed in these species, presumably because of their importance during a fight. Male combat is certainly not something which could be easily seen in the laboratory stick insect *Carausius morosus*, for nobody has ever actually found a male. This species reproduces exclusively by parthenogenesis, whereby unfertilized eggs give rise to generation after generation of females. Parthenogenetic reproduction is common among phasmids, but in most species it only comes into play as a back-up device if all else fails and no males turn up to effect fertilization. Parthenogenesis is also found in the Orthoptera in at least one grasshopper, three species of raphidophorids and three kinds of katydids.

Chapter 3

Egg-laying and Development

The function of mating is to fertilize the eggs contained within the female's body. The development and subsequent treatment of these eggs differs according to species. The majority of the orthopteroid insects abandon their eggs once they have been laid in a suitable spot. In some species maternal care is more advanced, leading to the guarding of the developing eggs inside or outside the body and finally to the evolution of family life in which the female and her offspring live together for some time. The immature stages of orthopteroid insects are usually called nymphs and are basically small, wingless versions of the parents. They develop to adulthood via a series of moults, each period between a moult being called an instar. The developmental series thus goes egg – nymph – adult, and is called incomplete metamorphosis. These nymphs are very different from the caterpillar of a butterfly or the maggot of a fly, whose larval structure is so radically different from their parents that they seem to be separate kinds of animals altogether. It is only after a quiescent

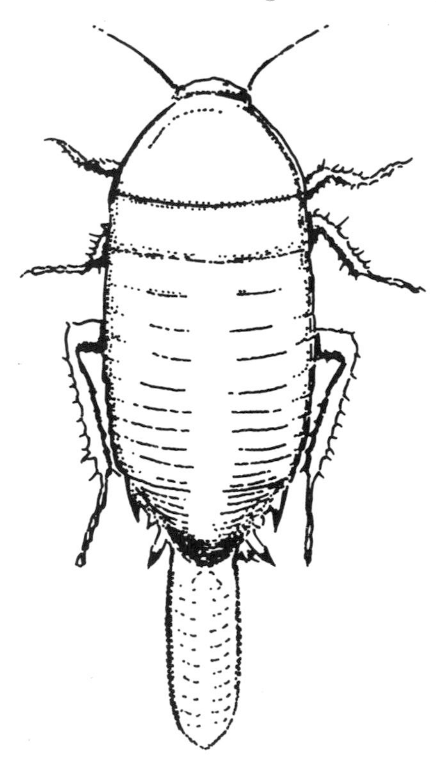

Fig.10 A female cockroach carrying her ootheca at the tip of her abdomen.

stage dedicated to complete structural reorganization, called the pupa, that caterpillars or maggots undergo the spectacular transformation to the adult state. In these insects the series therefore runs egg – larva – pupa – adult, and is called complete metamorphosis.

BLATTODEA

There are two main methods of reproduction within the cockroaches, both of which involve the production by the female of a specialized egg-capsule called an ootheca. In the first or oviparous method, this ootheca is usually composed of two rows of eggs surrounded by a protective envelope formed from glandular secretions. This is at first soft and pliant, but soon hardens to form a tough outer skin. In many species the ootheca may be a conspicuous object which remains attached to the tip of the female's abdomen and is carried around by her. Other species simply attach the ootheca to some suitable object, cover it with bits of debris to conceal it and then leave the eggs to their own devices.

In the second or ovoviviparous method, an ootheca is also produced: not the tough structure produced by oviparous species but a much smaller, softer, more flexible-walled affair which is formed externally and then rotated through 90° so that it can be retracted into the female's body. Here it lies within a special uterus or brood-sac until the eggs, which absorb water from the mother, are ready to hatch, when the ootheca is extruded. This is the standard method in the large family Blaberidae, in which eight of the nine subfamilies are ovoviviparous, but so far only a single member, *Symploce bimaculata,* of the large, mainly Old World family Blatellidae is known to exhibit similar habits. Ovoviviparous cockroaches produce fewer oothecae than oviparous kinds, but compensate for the discrepancy by multiplying the number of eggs per ootheca, which may sometimes be as many as 80.

A third reproductive strategy called vivipary also exists, but so far as is known it is restricted to a single species, *Diploptera punctata.* In this fascinating insect the process has advanced a stage further, for the eggs are particularly small and are not provided with enough yolk for complete development. To supplement this meagre built-in food supply, the growing embryos imbibe both water and nutrients, which are manufactured and transported by the female's brood-sac during gestation.

Some female cockroaches show quite a well-developed brooding behaviour to their newly hatched young. This 'hen and chicks' routine has so far been noted in around 15 species and has been extensively studied in a few. In others a great deal of mystery still surrounds the details of exactly how mother and offspring interact with one another. In certain species the relationship is both ephemeral and undeveloped. In *Nauphoeta cinerea,* the newly hatched nymphs seek protection beneath their mother's body, but only for a few hours until their rather soft and vulnerable cuticles have hardened sufficiently for them to risk making tracks into the outside world. The Cuban burrowing cockroach *Byrsotria fumigata* is an ovoviviparous species whose freshly emerged brood cluster closely around their mother. She connives in this by staying put and making herself available as a static shelter for reasonably long periods. The first instar nymphs recognize their mother's position and can distinguish her from other strange females by her emission of an aggregation pheromone. However, as the nymphs grow and moult into the second instar, their ties to their own mother grow weaker and they will happily court adoption by congregating beneath a neighbouring female. In this species, the female takes a rather detached view of her maternal duties, doing little except raise her body sufficiently for her offspring to crowd in beneath. But this is enough to boost her progeny's chances of survival beyond the first few highly vulnerable days.

Some cockroaches appear to have taken their post-natal chaperonage one step further and actually tote their babies around with them. The wingless female of *Phlebonotus pallens* from India and Sri Lanka possesses large, strongly convex wing-cases (tegmina) which sweep upwards over her abdomen, whose rather raised sides enclose a shallow depression, thereby forming a six-legged nursery-chamber for the young nymphs. Ensconced therein the babies are packed neatly away from the prying attentions of enemies such as ants, while in no way hindering the day-to-day routine of their mother. *Perisphaerus* females seem to carry their offspring slung in rows on their undersides. Females of *Perisphaerus semilunulatus* from the Far East resemble squat black pill-millipedes and roll up in a similar way into a protective ball which conceals all the more tender body-structures beneath an armoured shield. What then of the dozen or so tiny youngsters which the female may be carrying on her undersides? Do they get mangled as the close-fitting armour-plates meet together in the rolling process? Or do they simply fall off and get left behind in the rush? Apparently neither, for it seems that at least nine (and probably more) nymphs can remain safely in place as their mother rolls herself up. That said, there still remains one big puzzle about these nymphs. Their heads and mouthparts are modified in such a way as to make it seem likely that they 'plug in' to 'food-dispensers' – specialized orifices between the female's legs – although this has yet to be clarified in the living insect. The dependence on their mother for the early part of their lives – at least the first two instars – is fairly obvious in nymphs of this species, on account of their lack of functional eyes during this period. The first instar nymphs of one of the most primitive of cockroaches, *Cryptocercus punctulatus*, are also blind and even the adults have smaller eyes than is normal for cockroaches. Family groups of this species live in rotting wood which also forms their main food-supply (see Chapter 4). The most highly developed form of maternal care is probably exhibited by *Leucophaea maderae*, for the female stays with the nymphs for some time and accompanies them actively on nocturnal foraging trips for food.

MANTODEA

A single act of mating provides the female mantis with sufficient stored sperm to lay more than one batch of eggs. In the Indian desert species *Humbertiella similis*, for example, the female lays between six and ten batches of eggs without further sexual contact, each batch containing anything from 20 to 31 eggs. The last session of egg-laying may take place as many as 19 weeks after she received her supply of male sperms. The smallest single egg-batch known in mantids is around ten while some of the largest species can churn out 400 at a single sitting. A minority of female mantids lay their eggs in soil, but the

Plate 31 This large *Sphodromantis viridis* mantis is making her ootheca beneath the branches of a pine tree in daylight in Israel. Note how the cerci at the tip of her abdomen are used as combs to smooth the outer coat of the still-soft ootheca.

majority attach them to some object, sometimes in surprisingly conspicuous places. The eggs are not just dumped willy-nilly but are carefully placed inside a special form of weather-proof protective enclosure, the whole structure being called an ootheca.

Females often undertake the rather risky task of egg-laying under cover of darkness, but if sufficient cover is available they may carry it out in daylight. This was the case when the author chanced upon a large grey *Sphodromantis viridis* deeply absorbed in her procreative duties beneath the effective screen of a pine branch in Israel. The ootheca in this species is typical of the type which is often placed in exposed positions.

The lengthy procedure makes engrossing viewing as the female methodically weaves the top of her abdomen to and fro across the face of the ever-growing ootheca, expertly applying precise doses of froth as the base and sides of the structure begin to take shape. With preparations complete, she inserts two rows of eggs inside this froth, which rapidly hardens to form a durable spongy capsule. This tough outer envelope cocoons the eggs and helps protect them from enemies (but see Chapter 7) and the drying effects of the sun. Each egg is carefully placed topside uppermost in its own chamber, which is furnished with a special one-way valve at the top so that the spindly babies can emerge without a hazardous struggle. The openings for these valves can clearly be seen in two rows down the centre of the ootheca. After completing her lengthy and demanding labours, the mantis may well devote the ensuing few minutes to grooming the tip of her abdomen. During its involvement in a considerable amount of complex movement, her rearmost extremities may well have become contaminated with excess froth, which would harden and contaminate the body if not quickly removed. To assist in this important activity, at least in some (perhaps all) species, female mantids can reach the very tips of their abdomens with their mouthparts by turning their bodies into a complete circle. this contortion is facilitated by the flexible membranes

Plate 32 After making her ootheca, a female mantis will often devote a great deal of effort to grooming the tip of her abdomen. This is probably why this small *Acontista* species female from Trinidad is engaged in this activity. She is able to contort her body in this fashion because of the flexibility of the inter-segmental membranes.

Plate 33 A number of African mantids stand guard over noticeably long and narrow egg-cases. This female *Tarachodula pantherina* from Kenya has a pattern suggesting that she may be warningly coloured. Her presence may not therefore just serve to keep parasitic insects at bay, but might also succeed in protecting her eggs against attack by visually hunting birds.

between the body-segments. The importance of such grooming can be judged by the time – up to 15 minutes or more – and care which may be given over to it.

In those mantids which abandon their ootheca after attaching it to a tree, branch, post or rock, the ootheca tends to be short and fat, often whitish or pale brown in colour, with a considerable depth of spongy coating on the outside. However, not all mantids desert their handiwork; certain species play safe by staying put and standing over their developing brood as guardians. As mantids tend to be rather elongate, narrow-bodied insects, the shape of the ootheca in these instances is tailored to follow suit, otherwise the female could not mount any kind of effective guard with her body. Thus the ootheca of African species of *Galepsus* and *Tarachodula* which remain on sentry duty is rather long and narrow and the outer coat may be thinner, harder and less spongy. The female will attempt to ward off enemies, while her presence probably helps to reduce desiccation by shading the egg-containing sector of the ootheca. In the case of *T. pantherina*, with her conspicuous black-and-white-spotted legs and orange-banded abdomen, there must be a suspicion that the body of the female additionally functions as a warning notice protecting her eggs from vertebrate predators. (See Chapter 5 for notes on warning colours.) The exposed position which she may select for her ootheca (in contrast to the secluded spots usually chosen by the cryptic *Galepsus*) seems to add weight to the argument that *T. pantherina* may be warningly coloured on account of some kind of chemical protection. Not much is known about the effectiveness of these maternal sentinels, but their sharp-eyed presence is presumably valuable for keeping tiny parasitic wasps at bay, as well as for repelling solitary ants or predacious bugs which regularly suck the contents of insect eggs.

If things go well, about a month passes before the tiny nymphs poke their heads out of the ootheca via the network of exit-holes situated on the upper midline. Upon emerging into the air for the first time, each nymph looks rather like a miniature sausage neatly wrapped in a close-fitting membrane. About half-way out of its cell this strange-looking creature pauses for a while, waiting until the pressure of blood pumped into the head splits open the membrane, enabling it to be shrugged off. In some species the hatching process is slightly more complicated, for the newly emerged mantids, still wearing their nursery-cowls, dangle for a while on silken threads before they finally moult and fall to the earth. The whole hatching process is quite protracted and represents one of the most dangerous periods of the nymphs' lives. In their crowded, helpless state they are particularly vulnerable to marauding ants, which with their excellent lines of communication are well-suited to cash in on their discovery by calling up a legion of extra recruits to cart away the entire clutch of babies. The hatchlings usually assemble for a while on their ootheca before dispersing, although some species will reassemble on a neighbouring leaf or twig. At this early stage the nymphs often resemble ants, which probably has considerable protective value against visually-hunting predators such as birds, which seldom eat ants. After a moult or two this form of mimicry is lost and the nymphs begin to bear a closer likeness to their parents, although the wings (at least in those species which have them) exist only as gradually

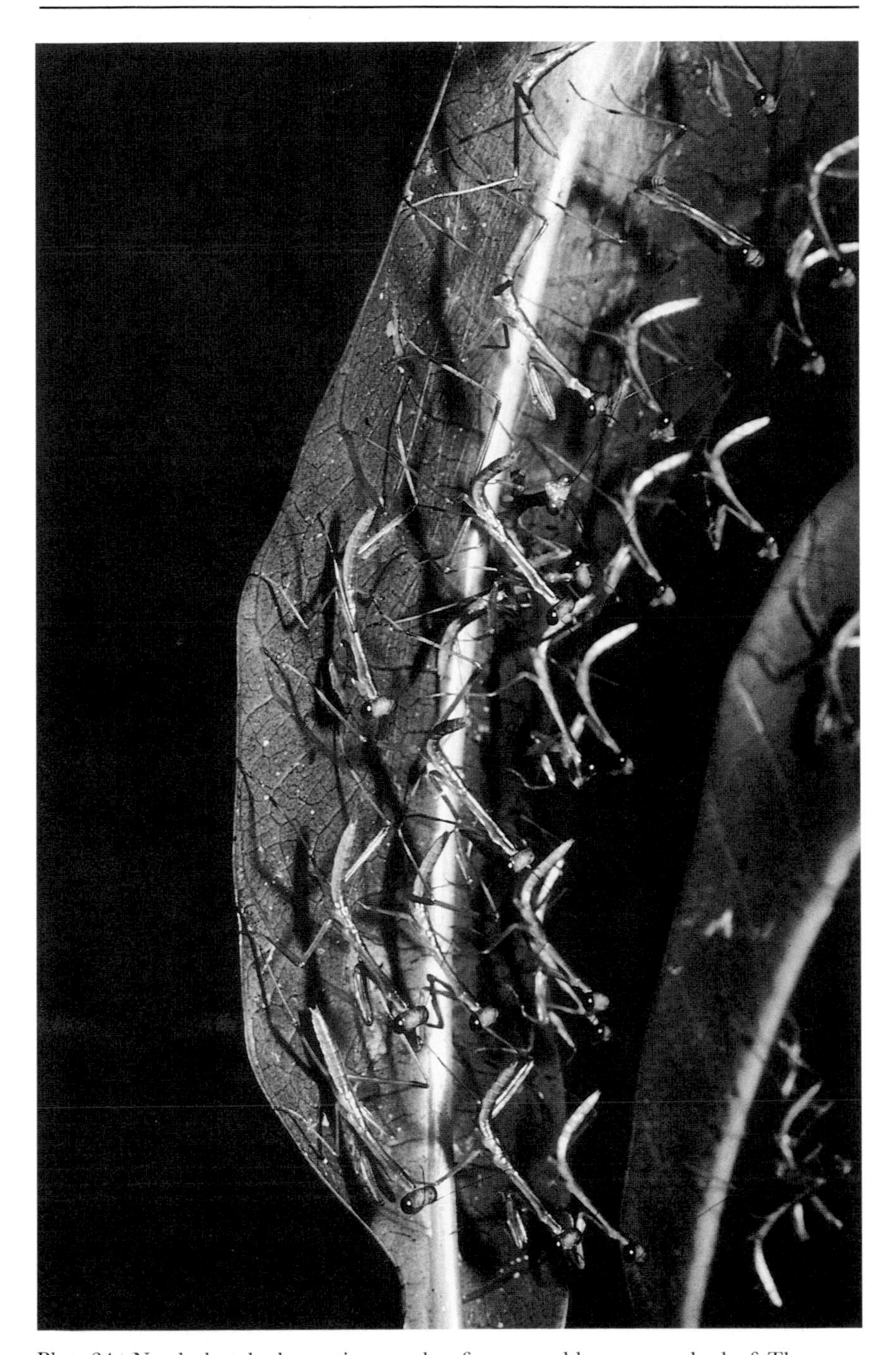

Plate 34 Newly-hatched mantis nymphs often assemble on a nearby leaf. These *Stagmatoptera septrionalis* nymphs are crowded on a leaf in rainforest in Trinidad. Unlike many newly-hatched mantids they do not resemble ants.

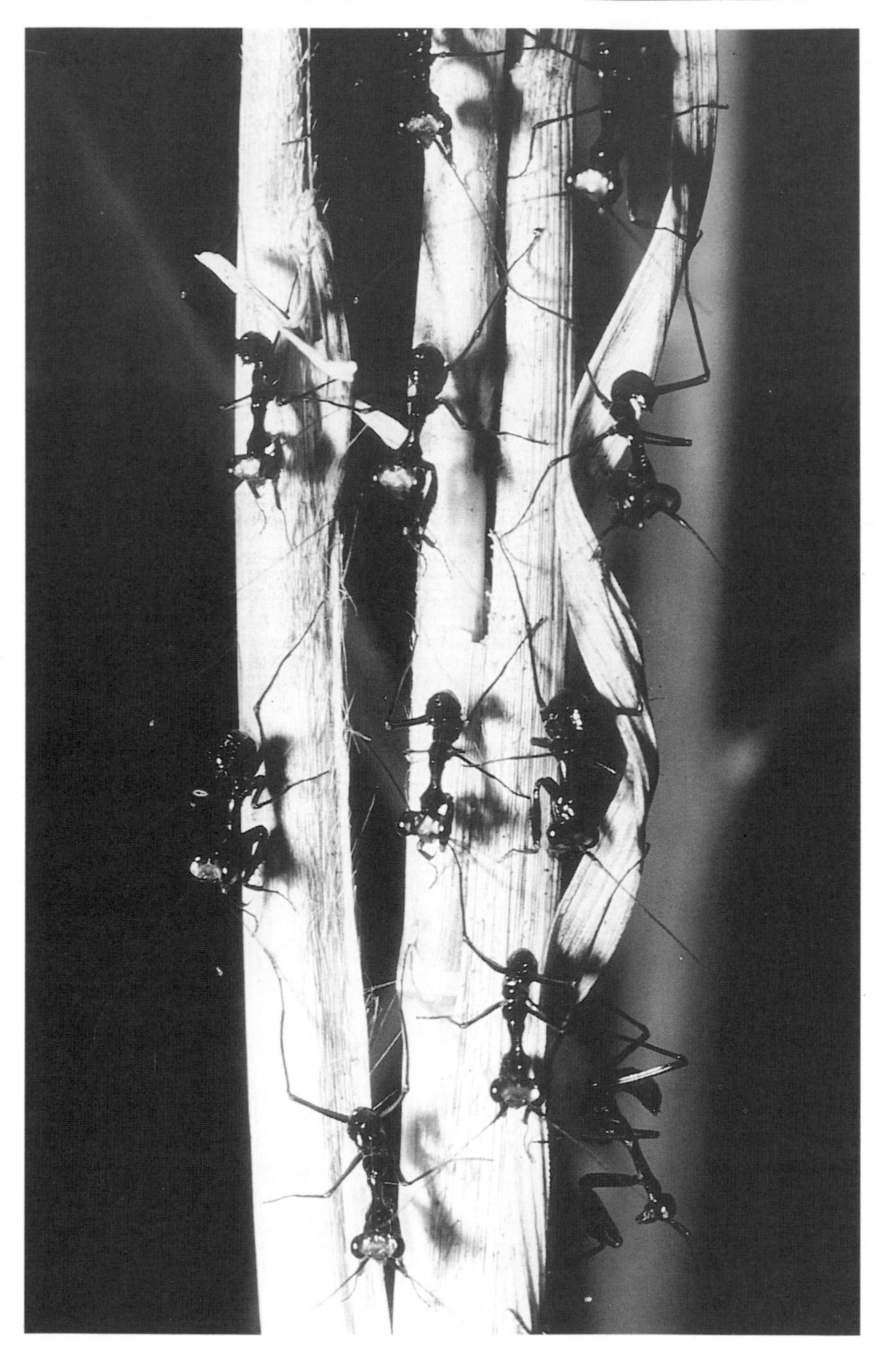

Plate 35 In their first two instars, mantis nymphs often mimic ants, a resemblance which is thought to give them considerable protection against visually-hunting predators, few of which take ants. These are from Kenya.

Plate 36 Female katydids generally have a very long sword-like ovipositor which protrudes conspicuously from the rear of the abdomen. This is a species of *Ephippiger* from France, a katydid which lays its eggs in the ground.

enlarging buds, until after the final moult. One particularly noticeable aspect of a mantis nymph is its habit of carrying its abdomen curled upwards over its back in a semi-circle, a highly characteristic but strange posture for which we currently have no explanation.

ORTHOPTERA

TETTIGONIIDAE

The most noteworthy feature on female katydids is the ovipositor protruding prominently from the end of the abdomen. In some species this is fairly short and broad, often curving upwards like a scimitar. Usually the ovipositor is longer – sometimes even exceeding the length of the rest of the body – and much straighter and sword-like. Some people treat these insects with considerable suspicion on account of this apparently threatening 'sting', but it is dedicated purely to egg-laying and is useless for defensive purposes.

Actually to come across a katydid laying her eggs under natural conditions seems to be rather a rare event, and despite thousands of hours spent in the field around the world, in every kind of habitat, both by day and by night, the author has only observed this event on a single occasion. This involved a European oak bush cricket *Meconema thalassinum* ovipositing into the lichen-covered base of a beech tree. Katydid eggs are generally rather sausage-shaped and laid singly, either in the ground or in vegetation; the method employed depends on which is chosen. Species which utilize the soil rear the tip of the abdomen upwards so that the ovipositor tip can point vertically and then push downwards into the earth, perhaps using the assistance of a natural crevice if this is available. Those species which lay their eggs inside the tissues of living plants usually pave the way by first biting a hole with the mandibles and then following up with the ovipositor. Only a single egg is laid in each hole, but in those kinds which exploit the cavity down the centre of grass or rush stems a large number may eventually be laid, filling the centre of the stem. In some species the tip of the ovipositor is furnished with a series of teeth which assist in sawing directly into plant tissues. Many tropical katydids lay their eggs in batches attached to leaves or blades of grass in surprisingly open situations. They are usually flattened and look very much like rows of melon seeds stood on end.

Katydid nymphs usually look more or less like miniature versions of their parents, although of course they never have wings. The nymphs go through a series of moults – usually six to ten – after each of which they resemble their parents a little more closely. The process of moulting is extremely hazardous as the insect has to remain for a prolonged period in a fixed position, able neither to leap nor even to drop to safety. For this reason moulting usually takes place under cover of darkness. At night in a tropical rainforest it can be quite easy to find katydids in the act of shedding their skins, although some nights seem to encourage this activity more than others; rather damp, overcast conditions seem to be favoured. The insects can be picked out very easily in the beam of a light, as the fresh cuticle is very pale and shows up clearly against the darker foliage.

Preparatory to moulting, the nymph's first action is to anchor itself firmly, usually to the underside of a leaf, using its tarsal claws as crampons. It then sucks in air in order to expand its body and force the beginnings of a rupture of the outer, unwanted cuticle. The initial stages of moulting are quite rapid, and the majority of the body is soon free. It now hangs backwards, secured by the tip of its abdomen which is lodged in the old skin, visible as a collapsed white husk still attached to the leaf. However, completion of the procedure is considerably delayed by the degree of care needed to extract the extremely long and flimsy antennae from their sheaths of obsolete cuticle. The insect executes this delicate feat very slowly, gradually inching the antennae out by pulling gently and steadily downwards until the tips finally come free. Even now, further specialized handling of these delicate accessories may be necessary, for the katydid's first action after the antennae are liberated is often to run them gingerly through the mouthparts at some length. The most perilous period is

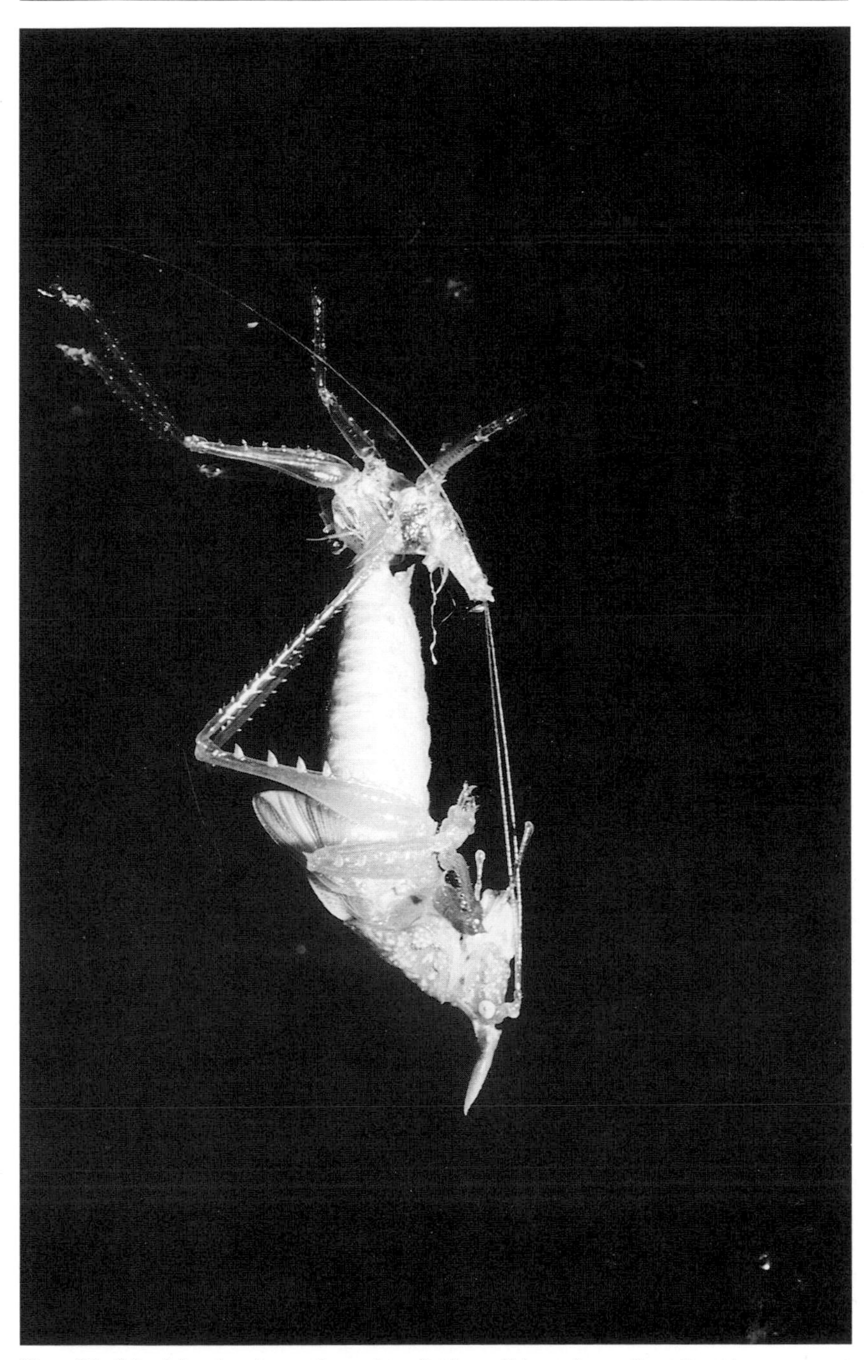

Plate 37 Moulting is a hazardous time for katydids and usually takes place at night. This *Copiphora rhinoceros* in a Costa Rican rainforest is gently extracting its long, hair-like antennae from their redundant sheaths.

Plate 38 After extracting its antennae from the old skin, a newly-moulted katydid will usually spend some time delicately grooming them. This unidentifiable species is moulting at night in a rainforest in Costa Rica.

Plate 39 After moulting is completed and their new skin has hardened, some katydids will proceed to make a meal of their old shed skin. This is *Ruspolia differens* in a rainforest in Kenya.

now over, for with its whole body free, the soft and pallid insect is able, if necessary, to move away to a place of greater safety. However, if undisturbed, it will generally choose to remain clinging to its old skin, although it now holds on with the two front pairs of legs, rather than hanging from the tip of its abdomen. If this is the final moult of a fully winged species, the wings will be clearly visibly as shrivelled little pointed buds projecting upwards over the pronotum. These rapidly expand into a set of pearly wings, although they are not functional until some time later. The final act may be to make a meal of the fragile scrap of cast skin, thereby recycling the nutrients contained therein.

Many species of katydids are wingless as adults and unable to fly. Colonization of fresh areas would thus be a relatively slow affair depending on crawling and hopping, a difficult exercise if blocks of unsuitable habitats intrude between favourable areas. The frequent solution is to provide fully winged adults if conditions indicate that migration would be a sensible thing to do. The usual stimulus to leave home is overcrowding, but not just any form of overcrowding will do; it must satisfy certain criteria. Thus in *Tessallana vittata*, normally wingless in the adult stages, the eventual production of winged forms is dictated by the degree of crowding at various stages of nymphal development. If crowding persists until the fourth instar, the nymph simply goes ahead and changes into a wingless adult, as normal. If, however, the overcrowding continues *beyond* the fourth instar, the end result is an increasing tendency to develop into a fully winged adult capable of relieving the pressure on food-supplies by taking off for fresh pastures.

GRYLLIDAE AND GRYLLOTALPIDAE

Female crickets also sport a conspicuous ovipositor which is used for laying eggs in the ground or in some suitable crevice, and in some species in living vegetation. The ovipositor is pushed vertically downwards, the penetration of the substrate being assisted by movements of the valves one against the other. Only a single egg is deposited, but as temperate species may go on churning out eggs throughout the summer, the eventual total can be quite large – up to 200 in the rather small European wood cricket *Nemobius sylvestris*. This species is of particular interest, for it has a biennial system of reproduction, with development from egg to adult taking two years. As a result of this, adults of one season rarely overlap with those of another, which would normally be too immature to be suitable as mates. So there would appear to be distinct strains of the wood cricket, even-year strains and odd-year strains. At present there is no evidence to suggest that these two strains are reproductively isolated, as has seemingly happened in the American cricket *Gryllus pennsylvanicus*. The populations of this cricket fall into two categories; those which mature in the autumn and those which mature in the spring. The songs of the two groups are indistinguishable and the only observable morphological difference is the length of the ovipositor in the autumn group, which is longer than in the spring population. There is a one-year life-cycle with the fall population spending the winter in the egg-stage, while the spring population does so as late instar nymphs. Final maturation thus occurs at different times, leaving little overlap during the six weeks of adult life for the two populations to meet and mate. In fact interbreeding never seems to occur in the wild and cannot be artificially induced, so the spring population has been given the status of a separate species, *G. veletis*.

The number of eggs laid by a cricket is sometimes quite large. The female of the cosmopolitan house cricket *Acheta domesticus* begins her reproductive career a week or two after her final moult into adulthood. She lays her eggs either singly or perhaps in small clusters. Final totals running into the high hundreds are normal for this species, but there are a few records of over a thousand. Temperature has an important influence on the time needed for hatching: perhaps two to three months at ordinary room temperatures, but only around a week in a really hot room reminiscent of tropical conditions. The number of instars is rather variable, from a low of seven to a peak of thirteen, although in Europe eleven seems to be the norm. As with hatching, development to the adult is much influenced by temperature, with five to eight months necessary at room temperature and only a month in hot, tropical-style surroundings.

Although the majority of female crickets abandon their eggs once they are laid, at least one species exhibits well-developed maternal care. This is *Anurogryllus muticus*, an inhabitant of the southeastern part of the USA. The female excavates an extensive subterranean burrow as a nursery for her eggs, which she aggressively protects against any intruders. The newly-hatched nymphs remain clustered near their mother, who acts as babysitter. She also frequently strokes her offspring with her mouthparts. From time to time she produces undersized eggs which her brood at once pounce on and devour. These eggs are apparently sterile and are seemingly adapted for their sole function of baby-food. An additional maternal task is to keep the burrow clean and prevent infection by regularly disposing of faecal matter.

Such maternal solicitude is perhaps more typical in the mole-crickets. In the European *Gryllotalpa gryllotalpa*, the female deposits her eggs in late spring in her underground nursery-chamber. She lacks the conspicuous external ovipositor typical of the preceding groups, and does not lay all her eggs in one go, but instead gradually builds up a clutch over a period of a week or two, much in the manner of many birds. However, no bird could exhibit such fecundity, for the female mole-cricket manages to lay 100–200 eggs, although more than 600 have been known. The parallel with avian behaviour continues, however, for mole-cricket mothers are diligent in their care for their eggs, which they guard against enemies. At regular intervals the female also licks the eggs, a habit shared with female earwigs, who also exhibit subterranean brood-care. It seems that the rather damp burrows tend to be excellent breeding-grounds for moulds, which quickly attack and ruin the eggs if the mother is prevented from licking them. Perhaps the licking removes fungal spores before they can germinate, although it is even possible that the female coats them with an active fungicidal secretion. The eggs respond well to the even temperatures typical of underground conditions, and hatch in two to three weeks. The hatchlings remain in the safety of their nursery for the first few weeks, subsisting on a diet of humus and tender young rootlets which protrude into the burrow. When

Plate 40 A large female *Lobosceliana* species grasshopper lays her eggs in a sandy path in a Kenyan savannah. When the author first spotted her he thought that a dead leaf was being pulled into a hole in the ground by some invisible animal beneath! Note the loose earth displaced by her extended abdomen and the absence of her left front leg. It is common to find Orthoptera with one or more legs missing, sometimes even both back legs. Even so, they seem to survive with few problems.

Plate 41 Grasshopper nymphs are basically small versions of their parents. They go through a series of moults, eventually assuming the adult coloration and finally developing wings (if these are present in the adult) at the final moult. These tiny first instar *Valangia* species nymphs are assembled on a leaf in rainforest in Malaysia.

Plate 42 The pattern in some grasshopper nymphs bears little resemblance to the adult coloration. This *Coscineuta coxalis* nymph from rainforest in Panama will shortly moult into an adult state which is green with bright red eyes!

Plate 43 A grasshopper's final moult may result in quite a drastic change in outward appearance. This *Taeniopoda auricornis* nymph started out the day as a shiny, black, wingless beast marked with yellow. It has just moulted into a pale brown and gold adult with large scarlet wings. At this early stage the adult colours are still not developed, while the wings can be seen as tightly folded buds just starting to unfurl. Several individuals were moulting in synchrony in the early morning in Veracruz State, Mexico.

they finally emerge into the open for the first time they do so under the protective shelter of their mother. They take up to two years to reach maturity and may attain a grand old age of more than three years.

ACRIDIDAE

All female grasshoppers lack the external sword-like ovipositor which is usually so conspicuous on female crickets and katydids. The eggs are generally laid in batches enclosed in a fairly durable case called a pod. This is normally placed in the ground, although in cool temperate areas, such as the British Isles, many species deposit their egg-pods among grass at ground level, while in tropical rainforests rotten logs may be the chosen spot. A few species are equipped for ovipositing directly into plant tissues, rather in the manner of katydids.

Oviposition often takes place under cover of darkness, but it is frequently possible to see a female performing this task during the day, especially in the tropics when the weather is cool and damp after rain. On such a day in the African savannahs it is quite common to find female grasshoppers of more than one species sitting in the middle of sandy tracks, absorbed in their egg-laying tasks. During her exertions, the female's abdomen may become completely buried in the earth, displacing excess material which is forced up into a mound around her. It is during this process that the female's abdomen shows what an extraordinarily flexible and telescopic organ it is. For example, the abdomen of an ovipositing migratory locust, *Locusta migratoria*, stretches from its normal length of 1 in (2.5 cm) to 3 in (8 cm) or more as she forces it into the earth, placing her eggs at the maximum depth, where security from enemies and desiccation is more assured. This amazing flexibility is facilitated by the extreme pliability of the inter-segmental membranes, allied to specialized muscles. The actual digging process is aided by four hard, horny points situated at the tip of the female's abdomen, which gradually penetrate the ground through steady thrusting and turning movements by the female.

The female *Phymateus morbillosus* of southern Africa penetrates the soil to a depth of $2\frac{1}{2}$–3 in (6–7 cm). She fills the bottom of the hole with a neat parcel of around 50 sausage-shaped eggs and then seals them in by venting a spurt of frothy liquid which soon hardens to form a durable envelope. She will repeat her actions two or three more times in different places until, with the coming of the cold days of autumn, she finally expires, leaving behind her legacy of around 200 eggs parcelled into three or four separate lots. In most tropical species the eggs hatch in three to four weeks, but in temperate kinds the eggs usually enter a period of suspended development or diapause to tide them through the winter months (in desert grasshoppers the diapause may take the eggs through the dry season). After several months, the eggs of *P. morbillosus* eventually hatch, revealing a wriggling creature bearing little resemblance to either of its parents. The first stage or vermiform larva is typical of grasshoppers, being well-suited to its job of making its way through the soil and up to the surface, a task which is more easily accomplished by a worm-like shape than one with tiny fragile legs sticking out awkwardly in several directions. This vermiform stage only lasts until the open air is reached and then the old skin is shed and a more grasshopper-like creature emerges.

The newly-emerged nymph is soft and pale, but its skin soon hardens and within a few hours it assumes its proper colour of shiny black. In this species the young hoppers lack the solitary inclinations exhibited by most grasshopper nymphs and instead stay together in prominent groups, moulting six times and going through several changes of colour and pattern until the adult stage is reached about three months later. The number of moults varies from species to species, while in many grasshoppers changes in colour after each moult are minor. Of particular interest is the way in which the moulting of gregarious grasshoppers is often closely synchronized, such that the nymphs of the African pyrgomorph *Phymateus purpurascens* will all shed their skins within a few hours of one another, at least in the first four instars.

Many tropical grasshoppers (at least those which are not forest-dwellers) choose open sunny ground in which to lay their eggs – the sandy paths already mentioned being a case in point – but a few seek out shady spots, as the eggs are susceptible to drying out and need moister ground than normal. The elegant grasshopper *Zonocerus variegatus* (Pyrgomorphinae) of West Africa is a good example, the females gathering together in favourable spots to form dense egg-laying aggregations. The place chosen is usually a shady patch underneath some woody plants, although it seems likely that in this species it may not be the females but the males who initially plump for any particular spot. Male gatherings soon build up, whose function is to attract the females, initially for mating and then for oviposition. These male assemblages form what is in effect an olfactory mating lek, for they emit a pheromone which is attractive to females. So the more males crammed into the concourse, the stronger the attraction and the greater the number of females who will be trawled from the surrounding areas by the odorous skeins of this wind-borne mating-net. These leks seem to be prominently successful, for large numbers of both sexes soon swell the ranks and the quantity of egg-pods deposited may reach several thousand, at a density exceeding 500 pods per square yard (550 per square metre). The lekking-sites can be remarkably enduring, with grasshoppers coming and going continuously for a fortnight or more. The females are regular but sporadic visitors, only calling in when they have an oviduct replete with mature eggs, but the males may stay longer and hang about in the vicinity, ready to launch themselves immediately on top of nubile females as soon as they put in an appearance. By the middle of the day, a popular site could well contain a concentration of several hundred grasshoppers, making an ostentatious spectacle, for this is an extremely flamboyant, warningly coloured species.

This is but one of many species of wingless grasshoppers which are capable of producing fully winged individuals under certain circumstances. Long-winged individuals in *Z. variegatus* are generally rare, probably because their maturation time is more prolonged. They are not therefore ready to lay eggs until after populations of a fly which parasitizes the egg-pods have become high. This reduces the chances of increasing the percentage of the gene responsible for producing long wings. However, the perennial survival of the long-winged form illustrates its usefulness in searching out new areas of suitable habitat. For this species, at least in Nigeria, this once consisted mainly of natural clearings in forest produced by tree-falls, although man-created habitats are now normally exploited instead. Production of fully winged forms in other flightless species is often a response to overcrowding, such as happens after a particularly favourable year for reproduction and survival.

A few grasshoppers have adopted the habit of laying their eggs on to or into the tissues of living plants. *Euthystira brachyptera*, a rather beautiful lime-green European grasshopper, attaches its eggs among grass-stems; this mode of attachment is termed epiphytic. Two Asian species, *Oxya velox* and *O. oryzivora*, affix their eggs to stems of rice when the paddy fields are flooded, while the site of attachment for a third species, *O. chinensis*, is between the rice stems and their leaf-sheaths. *Marellia remipes* from Uruguay sticks its eggs to the undersides of the floating leaves of water plants. In Canada, *Neopodismopsis abdominalis* utilizes the partly dried dung of cattle and moose for this purpose. *Leptysma marginicollis* from central Texas employs endophytic oviposition inside the stems of water plants such as the giant bulrush *Scirpus californicus*. The exact spot is not chosen until the female has made a thorough examination and evaluation of the stem with her antennae and mouthparts. Having made her choice she gnaws at the stem and chews out a small hole which extends into the central pith. This initial incision merely serves as a guide-hole, for now she moves up the stem and, inserting the tip of her abdomen, begins to enlarge the opening, using methodical movements of her ovipositor valves. Finally her abdomen is intruded deep within the stem and the eggs are laid, topped out with the usual foamy protective plug.

PHASMATODEA

After the trouble and dedication exhibited by females of many of the above insects in order to secure a safe home for their eggs, it seems strange that many stick insect mothers approach the task of reproduction imbued with some of the most throw-away and carefree attitudes to be found in insects. In a number of species the female merely drops her eggs at random as she feeds. However, this behaviour is not as perilously negligent as it sounds, for the degree of care needed in depositing a batch of eggs is basically linked to their ability to survive a variety of hazards. Stick insect eggs, especially those typical of the 'throw-away' brigade, are beautifully designed to survive the female's slipshod treatment. The eggs of the South African species *Macynia labiata* are typical of this group. The slender green females drop their eggs, one at a time, at a rate of around one a day. The egg is a fascinating structure, for it resembles nothing so much as a polished green seed, decorated at one end with a conspicuous white 'lid' which is detached by the baby stick insect when it emerges. The vital combination of hard shell and large size make the egg well-adapted for surviving in the open, or perhaps shaded by dead leaves or a bush, right through the sweltering African summer. Finally, after six months, the baby stick insect emerges and climbs on to a bush to feed. The chances of the egg remaining intact until now are increased by its excellent camouflage and the fact that the females of this and other 'throw-away' species are fairly mobile. They therefore tend to drop their eggs in widely separated places each time, making it an unprofitable endeavour for any predator to find more than one or two.

Species whose females tend to be rather sedentary achieve the same ends by catapulting the eggs some distance away from the food-plant. *Cyphocrania gigas* employs a method of ballistic oviposition which involves flicking her abdomen backwards, sending the egg flying a distance of some 20 ft (6 m). This type of oviposition is also typical of the stick insect *Extatosoma tiaratum* from Australia and the leaf insect *Phyllium bioculatum*. A female *E. tiaratum* lays around ten eggs per day, which adds up to a total of around 500 during her lifetime. She hangs

Plate 44 This *Damasippus* species stick insect is laying an egg in some old disused spider silk on a tree-trunk in rainforest in Trinidad. A number of phasmids attach their eggs to natural objects, while others simply broadcast them at random from their foodplant.

upside-down with her broad abdomen curved downwards. This is suddenly straightened and then instantly returned to its former position, thus projecting the egg several yards from the eucalyptus tree on which she is feeding. This method ensures that the eggs end up outside the zone directly beneath the tree where the female's frass falls. This is important, because a number of parasites and predators use the odour of frass for tracking down their target-species, and would more easily locate the eggs if they were living among a litter of easily traced droppings.

Other stick insects cement their eggs to some part of their surroundings, such as leaves, sticks, tree-bark, rocks etc. Depending on the species, they may either be deposited singly or in a pre-arranged pattern, for example in parallel lines. Egg-batches which are glued to green, living leaves in areas with a marked seasonality are primed with a fast hatching-time, so that the young can emerge before the leaf is shed from the tree. Stick insects which live in trees near water face a particular problem in ensuring that their eggs stay put and don't end up having a bath. *Megacrania* females which live on seaside *Pandanus* palms in Indonesia deposit their eggs safely within the palm's axils. Some species are flexible in their approach and may use more than one method. *Timema californica* lays eggs which at first have a lacquered and viscid exterior. These may merely be dropped to the ground, or else carefully cemented to the undersides of the leaves of the food-plant. Some species seek out crevices or some natural material in which to oviposit. In Trinidad, the author observed a female *Damasippus* laying eggs in an old silken cocoon whose original and now redundant function had been to protect a spider's eggs. Yet other species place single eggs in soil or sand, while *Anisomorpha buprestoides* from Florida displays a dedication rare in stick insects by going to the trouble of excavating a small hollow in the sand, which she digs out with her front two pairs of legs. Pausing with her head facing her handiwork she suddenly curls her flexible abdomen up over her head and neatly drops an egg into the hole. This procedure is repeated eight to ten times before she rakes the sand back over them for protection.

Mention has already been made in Chapter 2 of the ability to reproduce parthenogenetically which is such a feature of the phasmids. When this happens eggs may be either fertilized, giving rise to males and females in even numbers, or unfertilized, normally producing exclusively females, although males do occasionally crop up in some species. An interesting aspect of this is the geographical bias which is often found, such that races of *Bacillus rossius* in Italy reproduce sexually while the race resident in France is exclusively parthenogenetic.

Chapter 4
Food and Feeding

The diet of the insects dealt with in this book runs from the general to the particular. Most cockroaches will eat anything which is vaguely edible; many katydids take a mixture of vegetable and animal food, both living and dead; mantids are obligatory predators and will only accept live prey, while certain grasshoppers are finicky specialists and will refuse to touch anything but the leaves of their specific food-plant. However, that said, the generalist feeding behaviour of many grasshoppers has earned them a place among the most successful of all groups of insects.

BLATTODEA

The feeding habits of the pest cockroaches are mentioned in Chapter 8. Surprisingly enough, relatively little is known about the diet of the numerous species of 'wild' cockroaches. This is probably due to the difficulty of making detailed observations of creatures having mainly retiring, nocturnal habits. However, all three species of Britain's rather small, insignificant cockroaches can easily be seen in daylight, often perched on flowers, especially of the daisy family, where they seem to be feeding on the pollen. Many cockroaches, such as the unidentified species which the author observed as it emerged in some numbers at dusk in the Mexican desert and climbed into low bushes to feed on the berries, are partial to ripe fruits. The large brown mountain cockroach *Aptera cingulata* from South Africa has similar habits and has been seen feeding on the berries of parasitic *Cuscuta* dodder plants. Wingless desert cockroaches often actively patrol their habitat, carefully searching the ground for anything which turns up, including dead insects, fallen fruits or flower petals. Species such as the large pale South American *Blabarus giganteus* which inhabit caves used as bat-roosts encounter no such problems. All they have to do is wait around and the food drops in, showered on them from above by the bats, who are messy feeders and drop a considerable proportion of their food on to the floor. Add to this the constantly renewed and lavishly produced supply of nourishing bat-droppings and you have the makings of a cockroach gourmet's paradise which *B. giganteus* is well-suited to exploit. Cockroaches inhabiting tropical rainforests, where the majority of the world's species are found, forage widely and make the best of what is available on a daily or weekly basis. At certain of the wetter seasons, fungi can be counted upon to appear, particularly certain soft-bodied kinds which sprout from dead wood, an abundant resource in the forest. Seasonally available fruits, flowers and seeds are also taken, as well as the shedding bark of certain trees. Like most cockroaches, those of the rainforest are therefore opportunistic generalists, exhibiting an adaptable and catholic tendency which has almost certainly led to the overwhelming success of the cosmopolitan pest species, all of which probably originated in tropical forests.

Members of the primitive genus *Cryptocercus* feed on wood, a habit which they share with the termites of the order Isoptera, distant kin which were formerly also included in the Orthoptera. In order to digest their food, both termites and cockroaches need the assistance of minute symbiotic protozoans. These are packed into the gut where they perform the difficult task of breaking down the cellulose content of the wood. The larvae of *Cryptocercus* are born minus their vital complement of protozoan assistants, but they make good their deficiency by dining on faecal material issuing from the anus of the adults.

MANTODEA

Being day-active insects with fiercely predatory habits, mantids hold a certain fascination for human observers which has led to reasonable numbers of observations concerning their dietary intake. Having said this, it must be stressed that in the author's experience it is comparatively rare to come across a mantis in the wild while it is actually engaged in feeding. Nine times out of ten it will merely be sitting there patiently keeping an eye open for the arrival of its next meal. This rather sedentary 'wait-and-see' method of gleaning a living is typical of mantids, although there are a few more active species, such as the ground-living *Amantis reticulata* from Brunei which strenuously runs its prey down, even zig-zagging wildly to keep up with its intended victim.

Such exertions are, however, the exception rather than the rule for this group of specialist killers (mantids never eat vegetable matter) which are superbly designed for their role of surprise executioners. Loitering quietly, waiting for a victim to show up has certain advantages over active patrolling. The mantis itself is less liable to be spotted by its own numerous enemies, while the efficiency of the mantid's prey-capture mechanism ensures that anything which does trespass within its broad catching-zone is unlikely to be forfeited due to poorly developed technique. Mantids therefore rely on widely spaced meals which are often quite large and secured without the wasteful expenditure of much energy. The completely unexpected lunge at lightning speed from close range is also far more likely to be successful than trying to stalk and pursue prey which is well equipped to detect and evade a moving threat.

Mantids are well-equipped to carry out their vocation of deadly ambush. In most species the insides of the front legs are liberally furnished with rows of spines. The spines on the femur are all slightly tilted in one direction, those on the tibia in the other, so that when the two sections snap shut, the body of the prey is spiked on a kind of mobile gin-trap. The likelihood of successfully hauling in a victim from maximum range is increased by the positioning of two special spines on the outermost part of the tibia. These spines are very long and curved inwards, acting as grappling hooks as they are slung outwards over the victim's body, which can then be raked inwards with little prospect of escape.

Tradition holds that the waiting mantis invariably sits with its front legs ready for action, held up just below the head as if at prayer. This may be true most of the time, but many of the larger types seem to find the position irksome and often sit for long periods with the front legs resting on the substrate. They are quickly raised up to the capture position if something interesting turns up.

Plate 45 *Blaberus atropos* is one of the largest of all cockroaches. This one is feeding on fungi on a dead tree at night in rainforest in Trinidad.

If an insect of suitable size wanders into view, the mantis slowly turns its head to weigh up the possibilities. Then, as soon as the mantis judges the target to be within range, the front legs are raked forwards in a flashing strike which only lasts between ten and thirty milliseconds. This capture-lunge is normally performed with extreme precision and a victim is seldom given a chance to escape. The possibility of fluffing its attack is greatly reduced by the lack of any need

Plate 46 The front legs are a mantid's equivalent of a gin-trap and grappling-hook. This photograph taken in New Guinea shows clearly how the opposing rows of spines interlock when the tibia is drawn upwards against the femur, giving the prey little chance of escape.

for the mantis to waste precious time, and possibly give away its presence, by turning to face its prey before making the strike. The capture-lunge can be made accurately well to either side of the body's centre-line, for a mantis can generally strike in whichever direction (within reason) the head is pointing.

It is the close co-ordination between the head and legs which makes this possible. A mantid's head is usually triangular in shape, with the very efficient eyes set as far apart as possible, high up on the top corners. This gives the mantis the best possible degree of binocular vision, an important factor in being able to judge distances accurately. Co-ordination between the angle at which the head is inspecting the prey and the angle at which the legs should be directed in a strike is achieved by means of beds of sensory hairs situated on the neck. These are in contact with the back of the head on either side. The degree of articulation exhibited by a mantid's 'neck' is quite remarkable, particularly when compared with grasshoppers or katydids. A mantis can turn its head completely to either side and even a little way towards the rear, as well as upwards at a considerable angle. It keeps in touch with the position of its head via the hair-beds. As the head is turned to the right, it results in deformation of the hairs on that side, while at the same time reducing any effect on the hairs on the left side. The information is sent to the brain via nerve transmissions, so the mantis is precisely informed of the angle at which to set its legs during the capture-lunge. With the system functioning normally the success rate attains around 85%. This falls dramatically to only 20–30% when both hair-beds are artificially 'turned off'; while the blocking of the emissions from just a single bed leads to even greater problems. This is because the mantis is duped into believing that its head is turned further towards one side than is really the case, leading to a bungled strike which is much too far to one side of the intended target.

It is sometimes a puzzle how a solitary mantis, perched in one place for considerable periods, can actually make a decent living out of a lifestyle in which serendipity must play such a part in arranging for meals to turn up. This does not apply perhaps to those species which, like some wasps, stake out a position near some object which is likely to prove an attraction to insects, such as a flower. Many of the common kinds of fairly large green or grey mantids will lurk hopefully beneath a flower, looking fairly obvious despite their efforts at concealment, their raptorial legs poised ready to lance upwards as soon as some unsuspecting bee or butterfly drops by for a quick snack. However, the stance required to seize an insect when the mantis is *beneath* a flower and the intended victim is up on top is rather awkward, and mistakes are liable to be made. Certain mantids greatly improve the odds by sitting right up there among the petals, waiting to receive any arrivals directly in a welcoming embrace. These 'flower mantids' include various species of *Acontista* from South America, *Hymenopus coronatus* from Malaysia and Indonesia, and *Harpagomantis* and *Pseudocreobotra* from southern and eastern Africa. The nymphs of the last-named will sometimes sit on a bare stem, where their resemblance to a flower in its own right is so convincing that they do a good trade picking off insects which are attracted to the 'flower'. *Pseudocreobotra* nymphs possess amazing reactions and an impressive degree of dexterity, for they are able to snatch out of the air even the most nimble-winged and aerobatic of insects, such as bee flies (Bombyliidae), which are theoretically able to get out of a tight spot by their ability to fly backwards as rapidly as forwards. But what about the majority of mantis species which just sit around on leaves, twigs or bark, with no sign of anything special about the location selected? Perhaps the very fact that these mantids apparently manage to make a decent living is a clue to just how much general invertebrate traffic there is, coming and going across the vegetation.

In common with frogs and toads, mantids suffer from a peculiar blind spot concerning the detection and recognition of their prey, for there must be some kind of observable movement, otherwise there is no stimulus for a capture-response. The author has watched a cryptic bark-inhabiting *Liturgusa* mantis in Trinidad walk straight past a moth which was pressed closely against the bark. The legs of the unsuspecting mantis passed within a fraction of an inch of the outspread wings, yet luckily for the moth it didn't flinch, otherwise an unpleasant fate would probably have been assured.

Mantids are not particularly fussy about what they eat and will munch away, with every apparent sign of relish, on just about anything which comes their way. This may even include insects which share the mantids' predatory ways, such as assassin bugs, stealthy killers which, if given the opportunity (which they are usually not) are capable of instant retaliation with their

Plate 47 Many mantids lurk near a flower and leap on insects which arrive to feed. This small green *Tithrone* species mantis had spent the day sitting on a leaf beneath a pink garden flower in Trinidad. Its patience was rewarded by the arrival of a fly. The mantis has begun its meal by biting off its victim's head and is now tucking into the juicy thorax packed with nourishing flight-muscles.

poison-injecting hypodermic mouthparts. Many of these bugs contain unpleasant substances which are advertised to most predators via bright warning colours. However, although at least some of these assassin bugs seem acceptable both to spiders and to mantids, other warningly coloured insects prove just too revolting even for a hungry mantis. In fact it seems that mantids may even have a well-developed and unexpected ability to learn by their mistakes and avoid such unpleasant prey, an ability which is normally ascribed only to vertebrate predators, such as birds. Under experimental conditions, individuals of the Chinese mantid quickly learned to avoid eating

Plate 48 Even a hungry mantis will fail to recognize a potential prey-item in the absence of any kind of movement. This *Liturgusa* species nymph on a tree-trunk in rainforest in Trinidad failed to notice the moth pressed against the bark right in front of the mantid's face. The mantis almost brushed against the moth's wing as it walked past, but the moth sat tight and saved its own life. Both these insects are cryptically coloured for a life spent in the open on tree-trunks. They are part of a rich and diverse community of animals which live in this specialized habitat in tropical rainforests.

specimens of the warningly coloured milkweed bug *Oncopeltus fasciatus*. In the wild these pretty bugs lounge around in some numbers on milkweed flowers, openly flaunting their colours and making a temptingly easy target. After becoming ill from their first encounter with these bugs, the mantids subsequently refused harmless beetles painted with the bugs' characteristic pattern. The genuine bugs often proved so instantly repugnant anyway that they survived close encounters with the sacrifice of just a few peripheral bits of antennae or legs – apparently this preliminary sampling was enough to convince the mantids that tucking into the main course would be unwise.

However, insects which use warning colours to advertise their ability to sting, such as wasps, may find that their uniform does little to save them from the epicurean attentions of a hungry mantis. The European *Mantis religiosa* will sometimes lie in wait outside the burrows of solitary wasps and ambush them upon their return. This is potentially a particularly rewarding gambit, for the wasps will usually be ferrying another insect to serve as food for their young. The mantis ends up claiming two meals for the price of one. Presumably the mantids avoid getting stung during this seemingly risky enterprise by the speed of capture and the way in which the wasp's body is held sideways in the vice-like grip of the horny and relatively impermeable front legs.

When a mantis begins a meal it will, if possible, take the first few bites from the head or neck, which quickly wobbles clear, thus instantly cutting short any tendency to struggle. This may be an important consideration when the prey is larger than the mantis itself, perhaps armed with some means of retaliation, such as heavily spined legs or a sting. Guillotining the head also gives easy access to the juiciest bits – the nourishing tissues packed inside the thorax. At least death comes mercifully quickly when this happens. When it is the rear end which is tackled first the moment of final demise may be much longer in arriving, and it is then rather horrifying to watch the cold, clinical efficiency with which the mantis gnaws away the tissues from its still-living victim like a child with a candy bar. Mantids are thrifty eaters and generally manage to chew away at even tough legs and head-capsules, their jaws nibbling methodically away with a side-to-side action like miniature mincing-machines, processing the prey into tiny morsels small enough to pass down the rather narrow gullet.

While mantids' concept of acceptable fare generally consists of other invertebrates such as bugs, beetles, grasshoppers, crickets, other mantids, moths, butterflies, flies, caterpillars and spiders, they are not averse to taking vertebrate prey if the opportunity arises. When the author was in Australia he was informed that prolonged screaming sounds heard during the night were sometimes traced to large mantids which were eating frogs. That the frogs were very much alive while this was happening was obvious from their pitiful shrieks. This account is backed up by a number of definite records, such as the observation of *Archimantis latistylus* attacking and eating a golden bell frog *Litoria aurea*. *Heirodula werneri* is said to make something of a habit of preying upon small tree frogs such as *Litoria caerulea*. In fact this mantis seems to go in for tackling oversize vertebrate prey, for it has also been seen to capture a small bird, the brown honeycreeper *Lichmera indistincta*. However its gruesome conquest of a gecko *Hemidactylus frenatus* (a lizard which could itself include certain

Plate 49 The mantis *Sibylla pretiosa* spends its life resting cryptically on tree-trunks. It feeds on other smaller insects which also live in this specialized habitat, including, as in the photograph, the nymphal stages of its own species. It is possible that nymphs of this mantis find that their own adults are their worst enemies. The photograph was taken in rainforest in South Africa.

mantids in its diet) in a house in Darwin was observed in considerable detail, and is worth repeating here at length. Its meal kept the mantis fully occupied for 90 minutes of hard, unwavering mastication. Twenty minutes into the feast, the gecko's tail, including the bones, had been entirely consumed. The unfortunate creature was still alive, and continued thus for some considerable time, while the major portion of its flesh was delicately combed off its bones by the rhythmic pulsations of the mantid's jaws. At some indefinable point the poor lizard finally expired, only its head and spinal cord proving too tough for the mantis to tackle, so they were unceremoniously dumped.

Perhaps an even more grisly repast was enjoyed by a large specimen of the Chinese mantis, now a well-established alien in North America. This enterprising individual had managed the rather surprising capture of one of that continent's denizens, a white-footed deer-mouse *Peromyscus leucopus*. The mantis took the first few exploratory nibbles from the mouse's nose and then gradually worked its way backwards, cleaning up all attainable hair, bones, and tissue on the way. After ten minutes or so, the mantid's hard-working jaws penetrated through to the brain and the mouse shuddered and finally died. This seems to be the only definite record of a mantis eating a mammal. Once a meal is finished, mantids are noticeably fastidious about cleaning themselves up. They run their front legs slowly and carefully through their jaws, teasing away every last morsel of left-overs which could compromise the operating efficiency of the legs in their role of breadwinners-in-chief.

ORTHOPTERA

TETTIGONIIDAE

Considering how abundant they are, it is surprising that the author has yet to see any British species of katydid engaged in the act of feeding. This sad omission is despite spending thousands of hours in the field, often where large colonies of species such as the dark bush cricket *Pholidoptera griseoaptera* are present. It seems likely that in Britain at least all feeding is undertaken at night, and this general rule probably applies to the majority of the world's species. Not surprisingly, much of what we know of the diet of katydids is based on what they will readily accept in captivity, rather than what they have been seen eating in the wild. In captivity many species will happily gobble up

Plate 50 Katydids often feed on flowers. This unidentified Mexican species is feeding on the pollen-rich disk-florets of a flower of the daisy family. These are often selectively eaten in preference to other parts of the flower lacking pollen, although the outer ray-florets of this particular flower do seem to have come in for some attention from something, although possibly not this particular katydid. This is a female, as can be seen from her broad, upturned ovipositor.

Plate 51 In the tropics, garden flowers often suffer badly from the depredations of katydids. This small *Phaneroptera brevis* is eating an aster flower in a garden in Malaysia.

Plate 52 Katydids will often eat fruit. This unidentified nymph is guzzling away at night on a large over-ripe papaya fruit in rainforest in Trinidad.

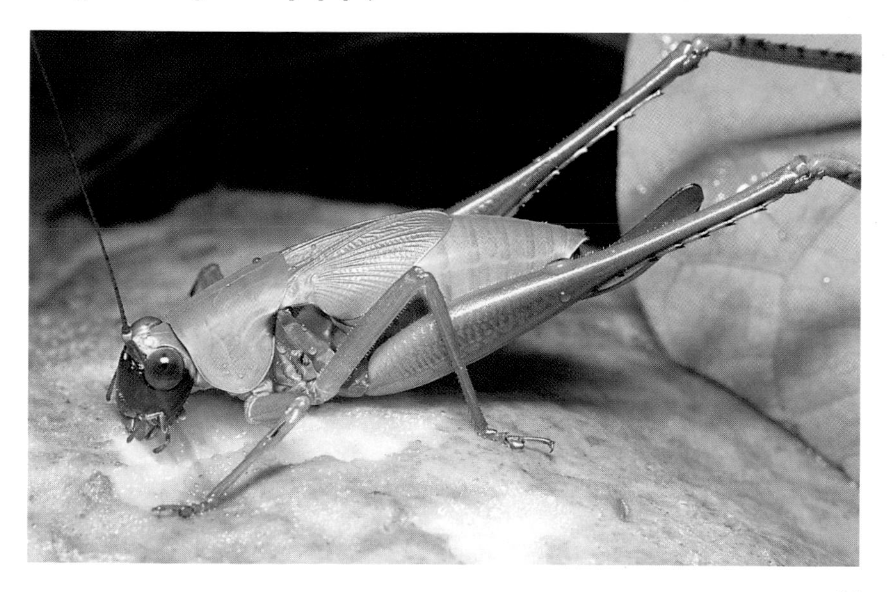

animal food with seeming gusto, mainly soft-bodied insects such as caterpillars. Despite this, it is doubtful if these constitute a major part of the diet for most species, being available more as a welcome dietary supplement on an opportunistic basis. The bulk of the intake is probably of vegetable matter. Although many species are probably exclusively vegetarian, the generalist theme probably applies widely, although there are exceptions, even among the tiny handful of British species. The oak bush cricket *Meconema thalassinum* seems to require animal food at all times. The same applies to certain tropical species such as *Copiphora* having powerful jaws which dictate careful handling, for they can administer a nasty and quite deep bite when picked up carelessly. Most members of the African Copiphorinae, however, use their competent jaws as nut-crackers to cleave open seeds. The cone-head *Pyrgocorypha hamata* has been seen voraciously attacking and demolishing tough-shelled chafer beetles (*Phyllophaga* sp.) attracted to lights at night in tropical forest in Chiapas State, Mexico. The large, rather ferocious-looking members of the subfamily Saginae are exclusively predacious and with their formidable jaws easily make mincemeat of other insects.

In the tropics and in desert areas it is often fairly easy to find the katydids feeding in daylight, although this is more often towards dusk than during the hotter parts of the day. Many species will sit and chew large holes in leaves, much in the manner of many grasshoppers, although flowers also seem to appeal to many katydids. Flower petals are often much less tough than leaves, while the anthers are covered with a candy-floss of pollen which is rich in protein. Often the whole flower is consumed, including the nectaries at the base which provide an energy-rich resource.

The flower is not always completely demolished, however, and indeed the katydid may prove itself to be a discerning connoisseur of its gastronomic qualities, showing considerable discrimination about which parts are eaten and which are ignored. A broad-winged bush katydid *Scudderia pistillata* seen feeding on a flower of the flat-topped white aster *Aster umbellatus* in Canada seemed to know exactly which parts to go for. Within the space of 20 minutes it munched away almost exclusively on the central disk-florets, which were mature and shedding liberal quantities of delectable pollen. Immature florets which were still closed were ignored, as were the conspicuous ray-florets, which have no pollen on offer, although one brief exploratory bite was taken out of one of these. This discriminating flower-fancier was thus able to tell the difference between the protein-rich, pollen-filled parts and the less nutritious reject parts lacking in pollen. Of course, this may simply have been because katydids become hooked on pollen as it tastes better than the rest of the flower. In some tropical countries this partiality to flowers can even make cultivation of certain kinds difficult. In Malaysia, the attractive pink asters frequently planted in gardens often suffer greatly from the depredations of small green *Phaneroptera brevis* katydids, while the crimson flowers of the widely planted ornamental *Hibiscus rosa-sinensis* often seems to suffer likewise all over the tropical world. Australian members of the subfamily Zaprochilinae seem to feed exclusively on flowers, making them unique among the tettigoniids. *Zaprochilus australis* will accept a wide range of flowers, including grasses and eucalypts, but seems particularly keen on the strongly scented white flowers of grass trees

Plate 53 Many katydids will take animal food on an opportunistic basis. This *Tettigonia cantans* was presented with a perfect target when a smaller species of katydid landed right in front of its face, having been disturbed by the feet of the approaching photographer. The larger katydid reacted instantly and trapped the smaller one in its strong front legs. *T. cantans* is a Central European species and was photographed in Switzerland.

Xanthorrhoea, uniquely Australian plants belonging to their own special family. Another Australian zaprochiline *Anthophiloptera dryas* is closely tied to the plant *Angophora floribunda*. The females lay their eggs in the bark and the flowers are the sole food-supply.

Katydids will also eat fruit. In the tropics quite large fruits may exert a considerable drawing-power, especially if they are somewhat overripe and beginning to disperse their aroma throughout the area like odoriferous magnets. In Trinidad a papaya fruit in this state provided a feast for several species of katydids, both adults and nymphs, which homed in on its succulent flesh under cover of darkness and guzzled away for long periods, their faces deep in the oozing pulp. However it was noticeable that when decay really began to set in the fruit suddenly lost its powers of attraction.

Another North American species, the shield-backed katydid *Atlanticus testaceus*, shows a definite preference for animal food but ends up with far more plant than animal material; animals are, after all, much harder to track down and catch than plants. It also shows a preference for dining on flowers, although leaves are certainly not shunned and fruits seem very acceptable. Animals which end up on the menu tend to be small, slow-moving and defenceless – in other words easy meat, such as small caterpillars, lace-wings, mirid plant-bugs, aphids and leaf-hoppers. Though also small, ants appear to be left severely alone, as are beetles whose armour is tough enough to resist the katydid's attempts to broach it. This katydid is, it seems, distinctly less intrepid than such obligatory predators as mantids, being rather reluctant to tackle larger prey which might bite back. If the opposition appears too hot to handle singly, several katydids might improve the odds by ganging up for a joint onslaught. Larger insects which for some reason are disabled or have their defensive capabilities rendered inoperative, such as when moulting, are fair game even for a lone katydid, and under these circumstances cannibalism seems to be a perfectly acceptable mode of behaviour. The actual techniques for taking small, defenceless prey such as aphids is hardly sophisticated – the katydid just dives straight in. But larger insects which are liable to ruin everything by flying off or running away are carefully stalked and then pinned down in a capture-jump, whose success seems to hinge on using the front legs to form a barbed stockade from which escape is difficult.

The African armoured ground crickets are also omnivorous, but show a distinct penchant for feeding on each other when the occasion permits. In parts of southern Africa large numbers of *Acanthoplus armativentris* are killed on the roads. Their mangled bodies prove an irresistible attraction to their brethren, which hustle out on to the road for their share of the easy pickings. Many of these quickly end up in the same flattened condition as their erstwhile meal, but new arrivals ensure that the gruesome feasting goes on and on in a self-replicating way until the corpses of armoured ground crickets litter the road in all directions in a clutter of crushed and shining debris.

GRYLLIDAE

In many ways the food preferences shown by crickets closely parallel those of katydids. Crickets are perhaps the most typical examples of thorough-going omnivores among the Orthoptera, taking a mixture of animal and vegetable food, not excluding the droppings of other animals, although again showing a definite inclination to include flowers as part of the diet, especially the pollen-rich anthers. Ground-dwelling crickets will climb up into low shrubs to feed on the flowers. Some species also consume large quantities of green leaves. The nymphs of a *Nisitrus* species cricket on Mount Kinabalu in Borneo regularly reduce the leaves of certain tropical forest plants to ragged skeletons. These crickets do not feed by straddling a leaf, as is common among grasshoppers, but sit on top of the upper surface and gnaw away holes between the leaf-veins, gradually moving forwards like a vacuum-cleaner, scooping up living tissue until the leaf is reduced to tatters. When crickets attack crops in this way they can attain the status of destructive pests. On the other hand, the northern field cricket *Gryllus pennsylvanicus* from the USA is considered a beneficial insect, as it regularly feeds on the crucifer flea beetle *Phyllotreta cruciferae* which is a major pest of mustard crops. As well as being cannibalistic when the chance arises, this cricket will also eat grasshopper eggs, moth pupae, caterpillars and insects which have been caught in spiders' webs. Pilfering from a spider's larder seems

Plate 54 Katydids will frequently stoop to cannibalism if circumstances permit. In nature this would normally mean when a conspecific is incapacitated during moulting, but in this instance large numbers of the armoured ground cricket *Acanthoplus armativentris* (Tettigoniidae) had been crushed when crossing a road in a dry area of Namibia. Their corpses attracted a continuous stream of new arrivals to a grisly repast. Many of these quickly succumbed to the same fate.

to be a risky way of getting a free meal, although it is regularly practised – albeit with great stealth – by scorpion flies of the order Mecoptera.

GRYLLOTALPIDAE

The same mixture of plant and animal matter also figures in the diets of

Plate 55 Crickets will often ascend the stems of plants to feed on the flowers. This *Cardiodactylus* species in Borneo is feeding on a *Lantana* flower on the edge of rainforest.

mole-crickets, although in some countries crops are attacked to such an extent that these insects are considered as pests. Feeding generally takes place within the subterranean confines of the insect's burrow, but this usually ramifies to an extent sufficient for the occupant to gain access to food over a surprisingly wide area. For this reason it is the roots of plants which are damaged, rather than the above-ground portions, but plants without roots don't grow too well, which explains the frequent pest status. However, mole-crickets also consume insect grubs, including those species which themselves damage roots, so it is not all one way. The African mole-cricket *Gryllotalpa africana* has developed a particularly interesting form of hoarding behaviour rare among insects. Germinating seeds of wheat and other plants are carefully gathered and stored in circular chambers below ground. In São Paulo state in Brazil, *Scapteriscus borellii* lives in a specialized riverine habitat where it consitutes a major predator of small subterranean ants, mainly belonging to the genus *Solenopsis*.

Raphidophoridae

Cave crickets are generally scavengers and will eat just about anything. In the cave itself fungi are often available, but the crickets are forced to come to the more brightly lit area near the cave-mouth to feed on green plants such as mosses and liverworts. However, there are various ways of effecting the delivery of extra rations from the outside world. If the cave has a river, this will transport quantities of plant material to the inhabitants deep within. In most parts of the world cave crickets also feed extensively on moths and butterflies hibernating within the caves, as well as adult caddis flies and other cave crickets. Where the caves are shared with rats or bats, their droppings may also provide an extra food-supply. Less is known about the dietary preferences of non-cavernicolous raphidophorids, although the New Zealand species seem to show rather an unusual preference for browsing on ferns.

In the USA more than two dozen species of sand-treader crickets are common members of the sand-dune communities spread widely over the southwest, where the crickets spend the day in simple burrows in the sand. A particularly large and impressive species of *Macrobaenetes* lives in dunes in southern California. Its normal diet is considered to be organic debris such as seeds, leaf-litter or dead insects, which are searched out during its nocturnal perambulations. However, these particular dunes are subject to periods of very high winds, which scour the sand clean and make feeding difficult. Under these conditions the crickets regularly resort to feeding on the droppings of lizards or kangaroo rats. To avoid having its discovery snatched from its jaws by the blustery wind, the cricket defies the weather by holding tightly on to the dropping in a corral formed of its own legs.

Acrididae

If you were to ask the average person to name the main food taken by grasshoppers you would probably land up with just a single answer – grass. The author admits to labouring under this common delusion as well, at least until he started travelling widely in the tropics. It was then that he discovered that many grasshoppers – perhaps even the majority – eat just about anything *but* grass. This inability to imagine that a grasshopper could possibly turn its nose up at a tasty blade of grass is due to having learned about insects from studying the British fauna. The mere 11 species of native British grasshoppers do include a few other herbs on the menu, but they are really just *hors-d'oeuvres* for the main course of grass, which certainly constitutes the overwhelming bulk of the diet.

However, to enjoy real success in a continent-sized country such as the USA, an insect needs to be more liberal in its perception of what is edible. A tendency towards over-choosiness and an inability to thrive on a wide range of plants will inevitably restrict the ability of a species to prosper and expand its range. Thus two of the most successful and widespread grasshoppers in North America are the two-striped grasshopper *Melanoplus bivittatus* and the red-legged grasshopper *M. femurrubrum*. Neither species is a finicky feeder and both will happily take a wide range of plants, although herbs are definitely preferred to grasses. Dandelions come out top, followed by chicory, red clover and plantains. Under experimental conditions the grass *Dactylis glomerata* seems to be acceptable as food, but amazingly enough nymphs which are fed solely on this soon die. Both species do, however, draw the line at touching one plant, the St John's wort *Hypericum perforatum*. This is not really surprising, as few insects other than certain specialists seem to be willing to tackle the plants in this widespread genus. Large numbers of plants including *Hypericum* and the majority of the tomato and potato family (the Solanaceae) produce poisonous or repugnant chemicals in their leaves which are designed to act as a deterrent to insect attack. While often proving effective against generalist herbivores, such as the two *Melanoplus*, these deterrents have in many cases failed so completely in their primary objective that some insects now treat the poisons as feeding *attractants*. For example, the pretty metallic *Drymophilacris* grasshoppers, which are a frequent sight in the forests of Costa Rica and other American countries, will only feed on plants of the Solanaceae, using specific chemicals resident in the leaves as a feeding stimulus. The link between the chemicals and the desire to feed is so powerful that in their absence the grasshoppers would rather starve to death than accept an alternative. Conversely, the grasshoppers will eat foreign substances as long as they are pre-treated with the necessary chemicals. As the poisons contained in the leaves of Solanaceae and other plants are often subsequently redeployed in the insects' own defences (but not, for some reason, in members of *Drymophilacris*), this whole subject is dealt with in more detail in the next chapter.

Their ability to capitalize on their catholic tastes has been a winning asset for the two *Melanoplus* grasshoppers, enabling them to become both widespread and successful in a great variety of habitats, for with so many food-plants being acceptable, the restrictions on spread are greatly reduced. This flexibility is in complete contrast to the two North American species of *Bootettix*, which are the only American grasshoppers known to be totally restricted to a single food-plant, the creosote bush *Larrea tridentata*. The distribution of both grasshoppers is therefore circumscribed by the distribution of their host-plant, which is common only within certain areas of the southwestern deserts. Yet even where the food-plant is abundant, *Bootettix* is not necessarily present. Of the two, the species more likely to be seen, as it can reach very high densities in favourable areas, is *Bootettix argentatus*, which lives among the leaves, where it is

Plate 56 The two American members of the genus *Bootettix* are the only grasshoppers in the United States known to be restricted to a single food-plant, the creosote bush *Larrea tridentata*. This is *B. argentatus*, by far the commoner of the two. This grasshopper always perches among the shiny leaves of its host-plant, blending in perfectly and virtually impossible to see until it moves.

very well camouflaged. It spends its whole life among the foliage and is reluctant to leave, usually merely fluttering quickly from one part of its bush to another when disturbed. The males call regularly but are not strongly territorial and several seem content to share the adequate food-resources encapsulated within their home bush. Such a tolerant attitude is notably absent in another grasshopper, *Ligurotettix coquilletti*, which also depends mainly, although not quite exclusively, on the creosote bush for both board and lodging. Browsing on the leaves only takes place under cover of darkness, for the foliage is no place for a slim grey insect such as this during the day, when it always clings to the stems, which the body outline and colour match closely. However, the males certainly make their presence known by singing virtually throughout the day in an emphatic acoustic assertion of their territorial rights. Intruding males are usually viciously attacked, although on occasion two males will be found sharing a bush in apparent harmony. Such active defence of a territory, even to the extent of resorting to actual physical assault is rare among grasshoppers, and this is the only North American species known to do it.

However, not just any creosote bush seems to merit the time and effort spent in its defence. Certain bushes seem to host a resident male year after year, while others are ignored. This contrast in popularity seems to be linked to the chemical composition of the resin which coats the leaves of the creosote bush. Resins containing a higher proportion of a particular acidic compound seem to be objectionable when consumed along with the leaf, so the males plump for and defend bushes which are low in these compounds. As well as providing him with a palatable food-supply, his selection of a consumer-friendly bush also maximizes his chances of making frequent sexual conquests. This is because the females are no fools either and are attracted to the better quality bushes for egg-laying purposes.

The creosote bush is also burdened with being the main food-supply for yet another grasshopper, *Cibolacris parviceps*, although unlike the other three, this species does not spend the day on its host-plant. Instead, it passes the daylight hours on the desert floor, where its general brownish cryptic coloration renders it extremely inconspicuous. Towards dusk the grasshoppers commute to the nearby bushes to commence the night's feeding. Mating takes place on the desert floor, so the males have no vested interest in setting up territories on the bushes.

A tropical rainforest would seem to be a strange habitat for such renowned sun-lovers as grasshoppers. Yet a surprising variety of species make their home in this shady, humid environment, especially in the forests of South America where members of the stick-like Proscopiidae lurk anonymously among the bands of brightly coloured, even jewel-like acridids and eumastacids. Each species of rainforest grasshopper tends to feed on a relatively small number of plant species, sometimes a single one, usually growing only up to a few feet above ground level. Wherever man has cut paths or roads through the forest, the extra light encourages a bonanza of lush rampant growth for the grasshoppers, which can usually be found in much larger numbers in such places than deeper in the undisturbed gloom of the high forest. Exceptions to this are certain ground-living grasshoppers which resemble dead leaves. The latter provide not only protective cover for their mimics, they also constitute the main diet for at least some of these grasshoppers, along with various kinds of fungi.

Most grasshoppers of temperate regions straddle the leaf as they feed. This position enables them to gain a secure purchase on either side of the leaf, with their front legs acting as a clamp while their jaws operate from side to side like a pair of clippers shaving away the leaf-edge. When feeding in this way the

Plate 57 Few tropical grasshoppers eat grass. These tiny newly-hatched grasshopper nymphs in a Trinidadian rainforest are feeding on the huge broad leaf of a *banana*. Their tiny mouthparts cannot bite right through the leaf-skeleton, but can only grow away the softer tissue from the upper surface.

grasshopper starts out by leaning slightly forwards, lowering its jaws until they touch the leaf-edge and then paring away a strip of tissue until the head is almost down between the front legs. It then stretches slightly further forwards and repeats the operation. When craning its head as far as possible fails to reach any fresh tissue, the grasshopper shuffles slightly forwards, or more likely changes its position completely to sample a different area of the leaf.

Rainforest grasshoppers seldom appear to adopt this procedure. Instead they almost invariably seem to sit on the leaf's flat upper surface and bite out holes by chewing downwards. This behaviour may be related to the generally much larger size of rainforest leaves, or perhaps to differences in the degree of concentrations of anti-predator chemicals contained in different parts of the leaf. Merely biting into a cassava leaf and rupturing the cells initiates a chemical reaction which releases quite large amounts of the extremely unpleasant poison hydrogen cyanide into the attacker's face. The oldest leaves pack a much stronger chemical punch, so insects such as the West African elegant

Plate 58 This small *Cephalotettix pilosus* grasshopper is feeding on an *Ipomoea* flower on a roadside through former rainforest (now sugar cane) in Mexico. It has used its front legs to pull over the edge of the flower so that its jaws can gain purchase on it. This is an example of the many kinds of small rainforest grasshoppers which are wingless as adults.

grasshopper *Zonocerus variegatus* often feed selectively on those parts of the plant whose guard is lowered, such as young tender leaves, diseased areas or older yellowing leaves which have lost their ability to gas their opponents. It is quite common to find rainforest grasshoppers chewing away on unappetizing-looking shrivelled brown edges to leaves, while leaving the toothsome-looking healthy parts in pristine condition, presumably for this reason.

Like katydids and crickets, grasshoppers also feed on flowers. On a roadside in tropical Veracruz in Mexico the author noted how the showy purple bell-shaped flowers of an *Ipomoea* were extensively pock-marked by numerous small wingless grasshoppers *Cephalotettix pilosus*, which also seemed partial to the anthers of a yellow 'daisy'. Under certain circumstances, flowers would seem to provide a virtual safety-net for the successful development of grasshopper nymphs faced with exceptionally hostile conditions. When the author was in

New Mexico early one May, the growth of the annual springtime flowers was extremely sparse following an unusually dry winter, and even the perennials showed little inclination to sprout fresh growth. In some areas the hedgehog cactus *Echinocereus fendleri* var. *fendleri*, with its ample food-reserves securely stored in its spiny succulent stems, had mounted an impressive display of self-sufficiency by shrugging off the drought and flowering prolifically. Indeed, the extravagant show was but little diminished from normal, more favourable years. Yet little of this floral endeavour seemed destined to pay reproductive benefits, for the anthers on the majority of the flowers over a wide area had been completely destroyed by *Melanoplus mexicanus* nymphs. These seem to have migrated to the flowers like bees to a honeypot, so that sometimes a single bloom fed and sheltered several nymphs. It seems likely that in a wetter year, when more abundant fresh growth is available, the cactus flowers might not be the focus of such plunder, although their particularly generous supplies of pollen probably always make them more attractive than most flowers.

Less respectable feeding-habits are displayed by certain rainforest grasshoppers which seem to find human excrement highly addictive. The forests near Tingo Maria in Peru are blessed with a particularly rich and colourful diver-

Plate 59 Some grasshoppers exhibit tastes which seem depraved to human beings. These brilliant *Paramastax nigra* eumastacids are feeding on the remains (now invisible) of human faeces in rainforest in Peru. The area is littered with grasshopper droppings, showing the intensity of feeding which has recently taken place on the unpleasant (to us) but nutrient-rich resource.

sity of grasshoppers (see Chapter 6). Some of the most abundant and conspicuous of these are members of the Eumastacidae. Several species are present, of which only one, *Paramastax nigra*, a gorgeous metallic green and gold creature with a blue head, exhibited any tendencies towards the unpleasant habits which hardly befit its regal appearance. As so often happens in the tropics, human excrement is often delivered in a hurry in an abnormally liquid form. It is frequently possible to find a number of these beautiful grasshoppers grouped in a circle around the periphery of such an odorous pool, their mouthparts buried in the mire and guzzling away with every indication of enjoyment. Occasionally they may be joined by single specimens of another grasshopper of similar tastes, the equally resplendent blue and orange *Cercoceracris viridicollis* (Acrididae: Romaleinae). Human (and other mammal) excrement is rich in nutrients and is utilized by many insects, particularly by the larval stages, but its regular exploitation by adult grasshoppers is probably fairly unusual.

PHASMATODEA

Stick and leaf insects are exclusively vegetarian, sitting motionless on plants – but not always their food-plant – during the day and moving around at night to feed. Most species are limited in their choice of food-plants, feeding on just one or perhaps a few species. Several Australian species which form plagues feed only on the leaves of certain eucalyptus trees. Other species are pests on coconuts in the Pacific Islands. As the greatest interest regarding their feeding habits lies in their defoliating capabilities, this group is treated more fully in the final chapter.

Chapter 5
Defence

The orthopteroid insects have a formidable array of enemies ranged against them. The pressure to survive in the face of such odds has given rise to a wide spectrum of adaptations. These fall into a number of broad categories which are explained below. However, it should be stressed that virtually all the survival mechanisms described in this chapter are aimed at protection from visually-hunting *vertebrate* animals, most especially birds, but also lizards and frogs, as well as mammals such as monkeys, civets, genets, mongooses, bats and rodents.

The following defensive categories are those most often employed:-

1. **Crypsis, otherwise known as camouflage**. At its most basic this may simply mean being the same colour as the immediate surroundings, as in the case with green mantids or katydids, which spend most of their time on green leaves, or the mottled or speckled patterns of grasshoppers which live on the desert floor or on sand dunes. This strategy, although it may be effective in general, has one great disadvantage. If the insect is forced to move to a non-corresponding background, or does so by mistake, then it immediately becomes obvious and may pay with its life. Thus the European speckled bush cricket *Leptophyes punctatissima* is a wingless green insect which, despite its rather plump body, manages to remain fairly inconspicuous when perched on vegetation. When it sits on a log or rock, it immediately assumes a prominence which makes it an easily recognized target. The insects in the next category do not suffer so markedly from this disadvantage because they have, in principle if not in fact, actually *become* a segment of their surroundings.

2. **Mimicking a part of the surroundings**. This category includes the plethora of species which mimic leaves of various kinds, as well as stones, twigs, flowers and bird-droppings. The only restrictive criterion is that the part of the surroundings mimicked should generally not be something which is regularly sought-after by insect-eating predators. Resembling such undesirable objects as dead leaves or bird-droppings would therefore seem to be a pretty safe bet. It also means that a katydid mimicking a dead leaf is not restricted to sitting among dead leaves on the forest floor. In fact this might be the worst spot to choose, as marauding ants tend to dominate this region, while the katydid would have to move to higher levels to find a supply of green food. But in reality dead leaves can turn up just about anywhere, so the katydid is free to sit where it likes without greatly increasing its chances of betraying its true identity. The same applies to mimics of twigs and even green leaves, both of which are regularly detached and can convincingly lie about in all sorts of positions.

3. **Warning or aposematic coloration**. Among insects, the habit of feeding on leaves which contain toxic chemicals is widespread. These poisons

originally evolved as part of the plants' defences against attack by their primary enemies, insects such as caterpillars, leaf-beetles and grasshoppers. However, not only have many insects become tolerant of these poisons, they have even taken up chemical warfare on their own account by sequestering the toxic compounds and re-deploying them as a mainstay of their own defensive armouries. Some insects, however, synthesize their own toxins from relatively innocuous plant precursors, and it is not always clear which method is used by any particular species.

Some of these chemicals are not actually toxic, but merely make the insect smell or taste so repulsive that no predator would want to swallow it, so that inducing rapid rejection, usually without any real harm being done, is the general aim. However, some insects, including grasshoppers, absorb chemicals such as cardenolides from their milkweed food-plants. These exert a violently unpleasant effect when swallowed by vertebrate animals such as birds. To prevent this happening in the first place, most chemically protected insects flaunt an easily memorized 'badge of office' permitting quick recognition before any harm is done. This 'badge' is usually some bright colour or colours, often applied in the form of distinctive striping or spotting. Some grasshoppers are among the most striking and beautiful of all warningly coloured insects, many of which form dense aggregations, especially in their immature stages (e.g. many pyrgomorph grasshoppers). This gregarious habit is well-founded. If a predator samples one or two members of a group and reacts by deciding never to bother again, the sacrifice of only one or two lives has been sufficient to protect the remainder. This apparently futile self-sacrifice is worthwhile because all the members of a group will have hatched from a single egg-mass, thus sharing many of their genes. Saving your brothers' and sisters' genes is preferable to losing the whole family, so the strategy of sacrificing the few for the good of the majority makes good sense.

4. **Mimicking other insects**. Warningly coloured insects, including those which are chemically protected or wield a nasty sting, are often copied by look-alikes or mimics which themselves are innocuous and would make a perfectly good meal. Such mimicry is not particularly common among the orthopteroids, but there are a number of outstanding and impressive examples, particularly among the katydids.

5. **Startle displays** tend to be used as a last resort to buy time. A palatable leaf-mimicking katydid which is pecked at may be able to save the day by a sudden startling transformation into something completely different, such as a fierce-looking, large-eyed animal which might pose a hazard to its attacker. The sudden flashing of eyespots or false 'faces', usually hidden beneath the wing-cases, is the most common method. Even chemically protected species may be generally cryptic, waiting until provoked before exposing some warningly coloured part of the body in an unmistakable 'keep off' message.

6. **Physical defences – striking back**. Some of the larger katydids, grasshoppers and mantids are equipped with strongly spined legs which are capable of drawing blood; this may be backed up by a painful bite.

Plate 60 Many warningly coloured grasshoppers aggregate in their early stages, especially in the Pyrgomorphinae, as seen here in a South African forest. This reduces the number of individuals which need to be sacrificed in order to protect the rest of the assemblage, usually consisting of brothers and sisters. The adult coloration is normally quite unlike that of the nymphs.

The well-developed back legs typical of many members of the Orthoptera are, of course, perfectly suited to catapulting their owner out of harm's way, while others may take to the wing as a primary defensive reaction. Orthopteroid insects also generally possess an excellent capacity to shed external appendages such as legs without coming to any harm, leaving their enemies with little to show for their endeavours, while the target makes good its escape.

Having explained the main defensive principles we can now consider their use in the various groups of orthopteroid insects.

BLATTODEA

Cockroaches are generally regarded as highly palatable and desirable fare by insectivorous predators. Their primarily nocturnal habits keep them out of the

Plate 61 *Homalopteryx laminata* is one of many ground-dwelling cockroaches which resemble the dead leaves among which they live. This is a wingless female, many of which fall prey to marauding army ants. Photographed in rainforest in Trinidad.

way of most enemies, which must seek them during the day when they are hidden away in crevices, inside hollow stems, under dead leaves and among general plant debris. Even when found, many cockroaches may prove difficult to pin down, being finely tuned to pick up vibrations of the substrate, to the point of being capable of detecting a movement of less than one-millionth of a millimetre. The abundant sensory-hairs situated on the paired cerci at the tip of the body also warn them of even slight air-movements set up by an approaching enemy, leaving the cockroach with ample time to slip quietly away into a suitable hideaway. Some kinds rely on their sprinting powers to get out of trouble, for cockroaches hold the insect record for fleetness of foot. Their long legs enable them to cover the ground at speed with remarkable ease, seeming to skate across a surface almost without touching it. Other types are far more slow-moving and depend on their cryptic appearance for protection. Species of *Homalopteryx* in the New World, and Old World species of *Rhabdoblatta* and *Gyna* live on the rainforest floor, where their flattened pale bodies closely resemble the dead leaves which surround them. This does nothing to save them from foraging hordes of army ants, and the wingless females have little choice but to stay put and submit to being torn apart. In the rainforests of Central America live several kinds of green cockroaches which sit around on leaves.

Plate 62 The South African cockroach *Aptera fusca* stands on its head with its 'tail' in the air and hisses menacingly when molested. This serves as a warning to beware of the cockroach's repellent defensive liquid and keep away. Its stance is similar to that of American *Eleodes* beetles which have similar defences, although these beetles cannot hiss.

A few cockroaches have a more active defensive system. The wingless American *Eurycotis floridana* (and other members of the genus) makes up for its inability to make a quick getaway by its rather skunk-like behaviour of emitting a deterrent, consisting of a smelly secretion from a gland situated between the sixth and seventh abdominal segments. The South African *Aptera cingulata* retaliates in a similar way, while *Aptera fusca* gives an advance warning of nasty things to follow by standing on its head with its tail pointed up in the air and hissing vigorously and very audibly. This defensive stance is strongly reminiscent of the well-known American *Eleodes* darkling beetles. Warning colours are relatively rare in this order, perhaps because a mainly omnivorous lifestyle is not suited to the amassing of the necessary chemicals. However, an unnamed species having the typical warning colours of a blue-black carapace ornamented with orange splashes was frequently noted by the author as it roamed around in daytime in forests in Madagascar. The author has also several times found dense rafts of 200 or more tiny cockroach nymphs on tree-trunks in Kenya and South Africa. These are only in place during the day – at night they disappear, presumably to forage. At close range these glossy blackish-brown nymphs are spotted with white and yellow, probably warning colours.

The habit of forming dense aggregations would tend to support this, but what these nymphs eat and how they would absorb any defensive compounds is unknown, unless they synthesize the necessary chemicals themselves.

The South American *Paratropes lycoides* has concave wing-cases marked with longitudinal bands of black, red and pale yellow. It sits in full view on top of rainforest leaves at low level and is said to be one of a whole group of insects which mimic unpalatable lycid beetles. The author has seen many of these mimics, which occur along with the cockroach, but should mention that his first reaction upon seeing this species was to mistake it for a small fallen fruit or petal, many of which litter the understorey. This may in fact be the primary line of defence, rather than lycid-mimicry. In the rainforests of Costa Rica, rather small, elongated, slim-bodied species of *Euphyllodromia* are commonly seen running over leaves in daytime. These are remarkably like some of the wasps which are abundant in these forests.

MANTODEA

Most mantids tend to be cryptic. Many of the larger, 'ordinary'-looking mantids can occur in more than one colour form, usually either green or light brown, but sometimes also grey. Generally speaking, each form tends to be adept at choosing a background which best matches its own colour, so that green mantids usually perch on green leaves, while brown forms elect to lie doggo among dead vegetation. The ratio of green to brown individuals in a population is not fixed, particularly in the African species. In these, green forms predominate in the wet season, while brown is more common during the protracted dry season. However, experimental evidence suggests that green forms are much better choosers of suitable backgrounds than brown ones, for reasons which are currently hypothetical. When it really matters, however, mantids may exhibit unerring skills in selecting a matching substrate. In Kenya the author found two female *Galepsus* guarding their oothecae. Each straw-coloured mantis had placed her like-coloured ootheca on the underside of a dead bleached doum palm leaf. The correlation between colour and shape of mantis, ootheca and leaf was so close that it was only after the most lengthy and painstaking searching that the two insects were discovered. These straw-coloured species of *Galepsus* and *Pyrgomantis* which inhabit African savannahs exhibit fire melanism, with black individuals predominating after a bush-fire; this also happens in some Australian species inhabiting eucalyptus forests. This interesting phenomenon is more widespread and better understood in grasshoppers, and is dealt with in greater detail below.

Despite the above comments, it is surprising how a mantis may survive perfectly well despite making no particular efforts at concealment. In Kenya, the author found a large grey and white female of *Polyspilota aeruginosa* perched in full view on a green leaf. In no way did she blend into her background and yet she was still there (within a few feet), unharmed, for at least a further six days. These are heavy-bodied, quite formidable insects whose fiercely spined front legs are capable of drawing blood when raked across a human hand. When threatened, their reaction is to rear upwards with their frong legs raised outwards to their sides, exposing a coloured area on their inner surfaces in a

Plate 63 These tiny cockroach nymphs aggregating on a tree in a South African forest would seem to be warningly coloured, a phenomenon which is rare in cockroaches.

most aggressive and ferocious-looking gesture of defiance. Many other large mantids adopt a similar threatening attitude when provoked and this sudden unveiling of coloured areas on the insides of the front legs is common.

Mimicry of some part of the surroundings, usually twigs or dead leaves, is widespread among tropical mantids. The best twig-mimics are the nymphs, whose anatomy and size enable them to make a much more convincing job of it than the adults. These nymphs often have a very rugose, bark-like cuticle which is convincing in itself, but their behaviour may reinforce the deception to a quite extraordinary degree. A nymph will hang head downwards with its front legs held straight out to the front. The tibiae are closed up against the femora, presenting a blunt end like a broken twig, while the antennae are concealed by being held back flush with the top of the prothorax. The abdomen is held upwards at an angle, as if the twig is bent, while the rear pairs of legs resemble dead shrivelled leaves or tiny twigs still attached to the stem. The whole effect is greatly enhanced by their skill at selecting a suitable background, for they have an almost uncanny habit of hanging up among twigs which they most closely resemble in colour, size and texture. The mantid's posture is therefore of considerable importance in maintaining its disguise, so it is vital that it should not blow its cover by moving. The importance of sticking tight was impressed upon the author when he spied a female *Catasigerpes* on a small tree in Kenya. This small mantis imitates a piece of shredding or peeling bark, and as there was plenty of the genuine article on this particular bush, she only gave away her presence by moving. The author took his eye off her for a second to adjust his camera, but when he looked back, she had disappeared. The ensuing ten minutes were devoted to a minute examination of the tree, finally resulting in the discovery of the quarry sitting immobile in almost precisely the same place as before. It was undoubtedly her immobility which had enabled her to remain undetected for so long.

A number of mantids spend their entire lives exposed on the bark of tree-trunks. They contrive to avoid the large numbers of birds which comb this habitat for food by having a dappled appearance which mingles well with the mosaic of lichens and mosses often decorating the bark. These mantids are also accomplished escape-artists, having the bewildering knack of seemingly vanishing in front of one's very eyes. This is due to their ability to dart sideways with extreme rapidity on their long legs, disappearing like magic around the other side of the trunk. The South American species of *Liturgusa* are prime exponents of this art. Similar but rather less cryptic mantids in the genus *Ciulfina* are common on bark in forests in tropical Australia, where *Ima fusca* lurks on the bleached boles of the paperbark trees, looking like a tiny wafer of the constantly shredding bark which drapes the trunks in untidy tatters. Some African bark-dwelling mantids are much larger, such as *Phloeomantis orientalis*, which matches the silver-grey trunks typical in tropical dry forests, and *Sibylla pretiosa*, which copies a fragment of peeling greenish bark. The nymphs of the latter species resemble tiny flecks of bark mounted on long thread-like stilts. They can run remarkably rapidly up the tree, where their worst enemies may be members of their own species, perhaps even their parents, who spend their lives hunting small insects on the very same trunks.

Some mantids, such as the African *Sphodromantis viridis*, bear a passable

Plate 64 *Phyllocrania illudens* from Madagascar is one of a number of mantids which are superb mimics of dead leaves. This is one of the most convincing of all. In particular note the outgrowths on the head, prothorax and legs.

resemblance to green leaves, but they lack the amazing perfection seen in many katydids or leaf insects. However, with dead leaves it is different matter, for when it comes to *being* a dead leaf, particulary the curled-up shrivelled kind, the prize for authenticity must go to mantids such as *Phyllocrania* from Africa and Madagascar or *Acanthops* from Central and South America. Members of the latter genus are particularly interesting because the males and females are strongly dimorphic and look completely unlike one another. The females are strange, wingless creatures whose habit of bending the abdomen upwards contributes considerably to their caricature of withered foliage. The effect is further enhanced by the rather distorted and dried-up appearance of the atrophied wing-cases, which taper off at their tips just like genuine leaves. The outline of the head is concealed between the flattened femora of the raised front legs. When hanging in her normal inverted position, the female's resemblance to a dead shrivelled leaf tangled up among a mass of living vegetation is extraordinary. If disturbed she may tremble slightly, as would a leaf in a wind, or else drop to the ground, where the task of locating her would be too daunting to be worthwhile.

In dire circumstances she can resort to another line of defence, namely display, which in various forms is also found in other cryptic mantids. Raising her forelegs and spreading them out on either side, she flicks up her wing-cases to reveal her small but brightly coloured yellow and black wings. In so doing she also uncovers her abdomen, which she twists to the side facing the aggressor, thus exposing more fully the abdomen's upper surface which is strikingly marked in rose-red and black. The formerly harmless 'dead leaf' is now instantly transformed into a fierce-looking and much larger insect which poses a flagrant (yet empty) threat by brandishing a warning type of pattern. Such displays are seldom employed by male *A. falcata*, which lack the female's bright 'flash' colours. The males have more flattened bodies and wing-cases which resemble a dead leaf in a far less shrivelled state. It is important to point out here that neither sex of this mantis lives among dead leaves on the ground, as would seem evident from certain photographs of this species which have appeared in books. These usually depict males which have been attracted to a light at night, subsequently being photographed, quite incorrectly, on the forest floor next day. In fact this mantis perches among living leaves, often low down on shrubs where the males may sit in full view on broad leaves (e.g. *Heliconia*). In such a position they are just one more dead leaf among the thousands which rain down throughout the year, never reaching the ground but adding instead to the jumble of detritus which accumulates on the mass of living vegetation at lower levels.

Defensive displays are not restricted to *A. falcata*. Some species have conspicuous eye-like markings on the insides of the front legs or on the wings. These are prominently flourished as the mantis rears up on its four back legs, trying to look as large and intimidating as possible. This may sometimes be assisted by a sinister hissing sound produced by scraping the upper part of the abdomen against the hind wings. These displays appear to be reasonably effective against lizards and small primates, although beating a hasty retreat always seems to be the last resort if things get too tough.

In the previous chapter there was mention of several species of mantids which live on flowers, which they may resemble to an amazing extent. Nymphs of the African *Pseudocreobotra wahlbergi* can change colour from white to yellow, pink, green or brown, depending on where they happen to be sitting. Although they may precisely match the colour of their chosen flower, this does not seem to be absolutely necessary, either for success in hunting or for survival. In Kenya, two bright pink nymphs spent at least six days perched in full and very obvious view on some small, deep blue flowers. Both nymphs were regularly seen eating small items of prey. One of the nymphs then moved to a stem bare of flowers and sat there, where its pink colour and petal-like adornments enabled it to pose as a flower in its own right. The masquerade proved to be of sufficient credibility to ensure a regular supply of insect visitors. The African devil flower, *Idolum diabolicum*, is said to hang head downwards beneath a branch, exposing the flower-like undersides of the prothorax and legs which are marked with dark dots resembling insects already feeding on the 'flower'. In South Africa, large numbers of the small *Harpagomantis discolor*

Plate 65 Mantids of the family Eremiaphilidae live on the ground in deserts in North Africa and the Middle East where they mimic stones. This female *Eremiaphila* species was discovered on open, sun-baked, stony 'reg' in Israel.

often station themselves on flower-heads consisting of numerous tiny bluish flowers surrounded by bundles of drab green bracts. The pink and green patterning of this mantis enables it to blend perfectly into such a background, while it could no doubt pose equally inconspicuously on a variety of other flowers.

Mimicking stones would seem an unlikely strategy for generally long-bodied insects such as mantids, but the short, squat members of the genus *Eremiaphila* of North Africa and the Middle East are remarkably good at being 'nomadic pebbles'. They inhabit the large expanses of parched, sun-baked, stony desert or 'reg' (the equivalent in the USA is desert pavement). Here, with little benefit of shade in the shimmering heat of midday, the mantids patrol the arid flats, their long spidery legs keeping their plump bodies from being roasted against the blistering stones. Equally hardy residents such as lizards and birds also mount their own food-patrols in this hostile landscape, where the mantids depend for survival on being just one more stone among billions. They are impossible to spot unless they move, and the author found that the only method of finding them was to pace rapidly up and down a predetermined area in the hope that a 'stone' would suddenly shoot out from beneath a descending foot and scuttle off. But its position must be firmly fixed, for if one's eyes wander for just a moment, the mantid's location will be lost.

In Chapter 3 it was pointed out that early instar nymphs of many species of mantids mimic ants. In Australia, the wingless adult females of an unnamed species and new genus also mimic ants. The mantids run in short, vigorous bursts and then stop and vibrate their antennae, just like the ants with which they run, generally on red lateritic soils. In some instances the mimicry may extend to copying ants of a particular genus or even species. In West Africa the mantis *Tarachodes afzelii* mimics *Camponotus acvapimensis*, *Miomantis aurea* mimics the brown weaver ant *Oecophylla longinoda* and *Miomantis paykulii* mimics species of *Pheidole*. In these cases it is generally the nymphs in the first one or two instars which do the mimicking, their ant-like appearance changing to something more normal as they become too large to be convincing as ants.

ORTHOPTERA

TETTIGONIIDAE

The main forms of defence seen in the katydids are crypsis and the mimicry of leaves, with the majority of species being green. Most of the more highly perfected examples of defensive adaptations are restricted to the tropical regions and the species found in temperate zones tend to be fairly unexciting. The European representatives more or less sum this up. Most species live at low levels rather than high up in trees. Many kinds are green and sit around on leaves looking reasonably inconspicuous. Several others are brown or grey and also sit around on leaves, often in large numbers, where they look much more conspicuous than the green ones. Presumably these rely on being exceptionally alert so that they can quickly dive for cover on the ground when disturbed. However, even some unimpressive-looking examples can pull something really special out of the bag when alarmed. The North American predacious *Neobarrettia* katydids are fairly average-looking, pale green insects which are by no means hard to spot when perched on leaves. If molested they react by suddenly holding up their front legs and flicking open their wings, which are vividly patterned with black and yellow. The wings are held in this position for as long as the threat is judged to persist, and then folded away again. Rather than being merely a 'startle' display this seems more akin to genuine warning coloration, possibly cautioning against the severe bite which this species is capable of inflicting, or perhaps signifying an unpleasant taste.

The tropics also abound with 'ordinary' katydids, not too dissimilar from their temperate relatives, and just as unimpressive. Unfortunately, the majority of katydids which the average searcher finds in a tropical forest fit this category, but there is always the stimulus of suddenly seeing through the disguise of one of the really fabulous leaf-mimics, an eventuality which never fails to keep the senses alert and the adrenalin flowing when searching for insects in the rainforest. And with good reason, for some of these leaf-mimicking katydids are impressive to a degree which leaves the observer lost in admiration and amazement. The many kinds which mimic green living leaves are wondrous in their perfection, but it is the species which faithfully copy dead or partly-dead leaves which are the real masters of the art of deception. The edges of the front pair of wings (tegmina) may be ragged, simulating the effects of decay or attack by leaf-eating insects. Silver specks mimic the effect of light shining through 'holes' in the leaf, while yellow or green blemishes mimic areas of incomplete decay. A leaf's typical mid-vein and the smaller veins radiating outwards to the edge are flawlessly mirrored by the insect's own wing venation. The deception is often completed by holding the back legs straight out to the rear, where they resemble leaf-stalks, while the front legs and head are tucked in tightly so as not to distort the leaf-like outline. Stance thus makes an important contribution towards perfecting the impression already established by colour and shape. In some cases a single katydid species, especially in Central and South America, can mimic leaves in various stages, from green living ones in unblemished condition, to green leaves with curled brown edges, yellow leaves with green blotches and brown leaves with green or yellow marks. The katydids cannot change from one type to another, this being fixed once the insect becomes adult.

For most of these katydids, the only hope of surviving discovery is to jump off their leaf and fall to the ground. Some species exhibit startle displays similar to those of the mantids, usually involving the sudden unveiling of the hind wings, on which two big coloured (e.g. blue and red) eye-spots are prominently displayed, so that the predator is suddenly confronted by a large and fearsome 'face' glaring at it from uncomfortably close quarters. Others have plain wings which are raised to expose a garish pattern of brilliant colours on the top of the abdomen; some do a headstand to show off bands of colour on the underside of the abdomen. The last two examples probably constitute genuine 'warning' displays in species which are chemically protected but normally rely on crypsis as a first line of defence. This probably also applies to the New Guinea katydids, notably in the genus *Sasima*, which stick their rear legs vertically in the air when molested and which mimic green leaves. These legs are carefully concealed beneath the body when the insect is at rest, but reveal striking areas of yellow and red when deployed in the defensive mode. Incidentally, as with the mantids, dead-leaf katydids are not generally found on the ground, but live on low-growing vegetation and in trees. Photographs depicting them on the forest-floor, scarcely visible in a sea of closely matching leaves, are generally carefully staged just for impact.

Tropical rainforests usually host diverse communities of animals resident on the trunks of the forest giants with their dappled coverings of mosses and lichens. In each of the world's main tropical regions there are several species of

Plate 66 Some kinds of katydid can occur in a variety of forms mimicking leaves in various stages. This *Typophyllum* species in a Venezuelan rainforest can be green, yellow or brown, mimicking leaves in different stages of decay. Even the brown individuals do not live among dead leaves on the ground but sit on vegetation higher up. This individual is sitting on a large *Heliconia* plant in which large quantities of the old dead leaves remain trapped for long periods.

Plate 67 Many katydids, especially those resident in tropical rainforests, mimic green leaves. This large species is *Sasima truncata* from New Guinea. The heavily spined border to the pronotum acts as a physical defence, while the red and yellow back legs are suddenly exposed and pointed skywards if the katydid is touched, indicating that it may be chemically protected.

Plate 68 The Australian *Acripeza reticulata* is one of a number of katydids which suddenly unveil brightly coloured parts of the body when molested. In this case it is the top of the abdomen which bears vivid blue and red bands, probably functioning as warning colours signifying distasteful properties in this slow-moving species.

katydids which are superbly adapted for life in such an exposed and apparently hazardous situation. Although they belong to quite different genera, they all share certain vital characteristics, determined by the need to remain undetected on the smooth, curved bole of a tree. They all have rather elongated, flattened bodies with the wings forming a tent whose lower edges are closely applied to the bark, thus eliminating give-away shadows. The long hind legs are held out straight at the rear, the much shorter middle legs are folded against the body, while the front legs and antennae are directed forwards. All parts of the body are pressed as closely as possible against the bark. The usual reaction upon discovery is to sprint as fast as possible around to the other side of the trunk, although species of *Acanthodis* from South America and *Cymatomera* from Africa may have recourse to a startle display of brightly coloured wings. In Peru up to three or four nymphs, belonging to more than one instar, of *Acanthodis aquilina* will often spend the day sitting close together on a trunk. If disturbed they all beat a hasty retreat in different directions, thoroughly confusing any predator which will be undecided which one to chase, thereby making it less likely that any of them will be caught. This species has extremely thick and sturdy back legs, which are jaggedly armed with rows of stiff spines. They kick out forcibly when alarmed, putting their legs to good effect as rakes to tear and scratch at anything which gets too close.

Such aggressive defensive tactics are commonplace in any large katydids having heavily armed legs, whose dissuasive powers may be backed up by a fearsome capacity to bite deeply and painfully. Both these abilities are amply possessed by the African species of *Clonia*, which have front legs like barbed wire and jaws like meat-cleavers. Their nymphs are less well-armed and rely on a completely different defence, for they mimic sticks and could easily be mistaken for a phasmid. The New Zealand *Deinacrida* giant wetas have enormous, stockily built back legs prominently furnished with sturdy spines. In an unmistakable gesture to keep clear the weta raises these vertically and makes sure the message gets through by stridulating. *Sasima spinosa* from New Guinea is also capable of brandishing legs full of spines in an aggressor's face, but reinforces the deterrent effect by butting with its heavily armoured thorax, which is fringed by a dense array of tough, pointed spikes.

The nymph of an unidentified species from Central and South America sits in full view on rainforest leaves, where it resembles a bird-dropping which has hit the leaf and splatted outwards. Several species mimic wasps of various kinds, modelling themselves on black and red braconids or the large spider-hunting *Pepsis* wasps with their tawny wings and blackish-blue bodies. Katydids which mimic such highly active, diurnal insects adopt the same habits and will move restlessly and jerkily through the forest as if searching for a host, exactly in the manner of the real wasps. The nymphs of some species mimic ants, although this is probably restricted to the first one or two instars before the nymphs become too large to be convincing.

Warning colours which are conspicuously visible at all times are rare among the katydids, even among those species which seem to possess at least a measure of chemical defence. The large, lumbering, armoured ground crickets which may be a prominent feature in deserts in South Africa can ooze a yellow fluid when molested, although their drab grey or brown colours hardly adver-

Plate 69 While they are in their early stages and still small, the nymphs of some katydids mimic ants. This tiny nymph is on a South African savannah.

tise the fact. Nevertheless, the defensive liquid does seem to relegate the katydid to somewhere near bottom spot on the menu for many predators, but if times are hard and food is in short supply, perhaps during a drought, a number of lizards, birds and mammals will eat them. The yellow mongoose

Plate 70 Around the tropical world there is a whole host of katydids adapted for a life on tree-trunks. There are two female nymphs of the large katydid *Acanthodis aquilina* closely pressed against this trunk at Tingo Maria in Peru. They are virtually invisible unless the trunk is very carefully searched. Note the heavy spination on their back legs which will be jabbed into an antagonist's face as a last resort.

Plate 71 The South African armoured ground cricket (a katydid) *Acanthoplus armativentris* is able to ooze a disgusting yellow defensive liquid, although it does not advertise this ability by flaunting warning colours. When prompted by extreme hunger some predators, such as the yellow mongoose *Cynictis penicillata*, will overcome their disgust and eat this katydid, despite its repugnant flavour.

Plate 72 Few katydids exhibit overall conspicuous warning colours, although many otherwise cryptic species expose warningly coloured parts of the body when molested. An exception may be this attractive *Dioncomena ornata*, which sits around conspicuously on leaves in daytime in Kenya forests.

Cynictis penicillata will pick them up and dispose of them to the accompaniment of much noisy crunching. The sound-effects are produced by the destruction of the thorax, which is heavily armoured and bears rows of rigid, sharply projecting spines around the sides and top. However, the explicit signs of disgust written across the mongoose's face make it obvious exactly how it rates its meal, which would probably be ignored in better times. When faced with predators other than the mongoose, with its powerful jaws and flexible attitude to flavours, the katydids' combination of armour, spines and disgusting fluid would probably be sufficiently potent to keep it out of trouble.

The author has seen only one katydid which could be classified as warningly coloured, the small African species *Dionconema ornata*. Its pattern of pale blue, black and greenish-yellow, allied to its habit of making itself look obvious by sitting around conspicuously on leaves in daylight, suggests that it may be chemically protected. A warning sign conveying another category of 'keep off' message probably explains the bright yellow spot on the face of a Trinidadian *Neoconocephalus* which, with its jaws like guillotines, is capable of administering instant painful deterrence. Splashing the warning sign directly above the instruments of retribution – the jaws – is presumably a particularly effective way of ramming the message home first time. The yellow horn sprouting from the head of the sharp-jawed 'unicorn' katydid *Copiphora rhinoceros* may serve a similar function, but the horn itself could inflict some damage in its own right, assisted by the heavily spined legs which are easily able to draw blood from a human hand or damage the delicate membrane of an attacking bat's wings.

One of the most unusual and unexpected forms of defence has been observed in the Costa Rican katydid *Ancictrocercus inficitus*. This species roosts during the day close to the nests of various wasps. Up to 16 katydids have been counted near a single nest, all remaining remarkably faithful to their adopted body-guards, returning day after day to the same nest after spending the night out on foraging trips. Trying to get at one of the katydids without upsetting their protectors is extremely difficult. The wasps swarm out of the nest at the slightest commotion, which itself may be deliberately provoked by a disturbed katydid jumping first on to the nest to alert its guardians, before falling to the ground out of harm's way.

GRYLLIDAE

Most crickets are fairly plump insects, drably garbed in shades of brown or black, and spending much of their time concealed from view. They rely on being able to escape from enemies by jumping or running quickly to shelter, and there are but few examples of the fascinating survival-aids seen in the katydids or grasshoppers. A few species bear specialized cryptic colours, such as are needed to survive in the open on tree trunks. The bodies of these bark-living crickets are generally rather flattened, as a deep body-section would throw a long shadow, giving away the insect's position. Rather than being brown or black, they are speckled in greys, greens and silvers, blending into the lichen-speckled trunk. In common with the bark-dwelling mantids, with which they may live, these crickets have a wonderful ability to dart smoothly across the trunk with amazing speed when disturbed, such that they seem to disappear from beneath one's mystified gaze.

A number of crickets seem to have warning coloration and spend the day sitting prominently around in full view. On Mount Kinabalu in Borneo, a species of *Nisitrus* is extremely abundant on vegetation along the roadsides. It is basically brown with two pale yellow stripes along the sides of the back and a network of yellow veins on the forewings. Groups of these attractive insects lounge around all day on leaves near ground level and take little notice of intruders, so it seems logical to assume that they are chemically defended in some way, a theory which is backed up by their habit of feeding on leaves. The same comments apply to the small yet spectacular *Rhicnogryllus lepidus* from rainforests in Kenya. This jewel among crickets is brilliantly attired in glossy brown, elegantly decorated with enamel blue markings. It poses very visibly in daytime on leaves in the forest's understorey. As in many mantids and some katydids, the young nymphs of some crickets mimic ants and run around on leaves or on the ground.

ACRIDIDAE

Photographing the generally dull and unspectacular ground-dwelling grasshoppers can be hard on the knees and soon impresses upon the mind the primary defensive philosophy of these insects – namely 'always keep your rear end facing the enemy so as to be ready to jump clear if it gets too close'. As pictures of grasshoppers' backsides obstinately turned towards the lens are not in great demand, it can be a relief to get to the tropics, where some of the gaudiest grasshoppers, resplendent in a coat of many colours, just sit nonchalantly and ignore the approaching photographer, instinctively trusting in their warning attire to keep them out of trouble. Between these extremes is a whole range of fascinating adaptations designed to increase the chances of staying alive.

In temperate areas most grasshoppers are cryptically coloured. Those which live in rather open, broken habitats such as chalk grasslands, heaths, savannahs or mountainsides, where patches of light or dark earth and stone intermingle with the varied greens, browns and buffs of the stunted vegetation to form a complex patchwork, tend to be shades of brown, green or grey, although the basic ground-colour is often fractionalized by a series of paler markings. The number of permutations is large, especially in a species such as the European mottled grasshopper *Myrmeleotettix maculatus* in which it can be difficult to find two specimens with precisely the same colour-scheme. It is noticeable that some individuals match their surroundings far better than others. It would seem logical to assume that natural selection would long ago have eliminated all patterns other than those which best merge into the background, but it seems that there may be selection pressure from predators, especially birds, to retain a large measure of variation. Birds are notoriously conservative in their choice of prey and tend to select food which most closely resembles something which they've previously eaten and found to be good – and the last meal down obviously supplies the strongest search-image for the next one. Grasshoppers which live in close association, such as the mottled grasshopper, would be at special risk from the first-caught of their number becoming the model for an avian 'wanted poster', enabling the whole lot to be

Plate 73 A gem among crickets is the small but beautiful *Rhicnogryllus lepidus* pictured here in a Kenyan rainforest. Its brilliant appearance would suggest warning coloration, something which is rare in the Gryllidae, most of which are sombre-coloured brown or black insects.

quickly identified and picked off. By all looking slightly different, these multi-patterned grasshoppers cash in on this conservative trend, as well as making it less likely that any predator can form a fixed search-image which would simplify the generally tricky task of distinguishing the insects from their surroundings.

The multiplicity of broken-patterned green species such as the European stripe-winged grasshopper *Stenobothrus lineatus* are fairly difficult to spot when lurking among grass, blending into the confusing muddle of criss-crossing blades inter-mixed with last year's pale dead stems. But they are still basically of a 'standard' grasshopper shape, and their rather stubby bodies only appear inconspicuous when seen from certain angles. This is not so in numerous species from around the world which have come to resemble the actual blades of grass themselves. These grasshoppers normally make an effort to orientate themselves vertically up a stem, but this is not strictly essential to maintain anonymity when the whole insect resembles a piece of grass. The head is generally long and narrow and protrudes a long way forwards, the flattened antennae are pressed down against the grass when at rest and the long, narrow wing-cases extend backwards way past the tip of the abdomen. The whole insect therefore has an elongated, low, sleek profile, enabling it to fit snugly against a stem without casting a shadow. Some species are quite large, such as

Plate 74 Numerous species of grasshoppers are both shaped and coloured to mimic grass-stems. This *Cylindrotettix* species from the Campo Cerrado in Brazil occurs both in straw-coloured and green forms, depending on the season. It is one of the best of all grass-mimics and is virtually impossible to see unless it moves.

Plate 75 The nymphs of *Prionosthenus galericulatus* from Israel resemble pieces of flaking bark or dead leaves. After fires the predominant colour among the nymphs is black, probably indicating an ability to change colour known as fire melanism, which is better known in certain African grasshoppers.

the European *Acrida hungarica* and many rather similar-looking African species of *Truxalis*. Others are smaller and slimmer, such as species of *Achurum* from Mexico and South American *Cylindrotettix*. An unidentified species of the latter genus is common in the *campo cerrado* near Brasilia. At the end of the dry season, when the grass is bleached and dead, the grasshoppers are mainly straw-coloured and difficult to detect. A week or two after the first rains, when the drab landscape has been transformed by a bright covering of fresh green growth, almost all the grasshoppers are also green. The rapidity of this switch seems to indicate that the grasshoppers have actually changed colour in order to suit the revised circumstances.

This colour-plasticity is well known in many African savannah grasshoppers, whose green or straw-coloured tints serve them well throughout most of the years. The triggers for the change are the fires which rage across the savannahs at the end of the dry season, converting the landscape into a ravaged wasteland of charcoal black and ashy grey. Within a short time of the passage of the flames, most of the surviving grasshoppers will have revamped their appearance and turned black, thus conforming with the cataclysmic change in their immediate surroundings. Not all the savannah grasshoppers are capable of this variation, the prime exponent being *Phorenula werneriana*, although at least six other species are also probably capable of a complete colour change. However, this fire melanism is apparently not restricted to the African savannahs. Fires are such a regular and ancient feature of Australia's eucalyptus forests that many plants will only flower after a fire has destroyed competing vegetation, while the searing flames may be the only means of opening certain fruits. Under these circumstances, it is not surprising that fire melanism has evolved in some of the grasshoppers, such as species of *Coryphystes*.

Even in Israel, the author found an example of fire melanism on the Mediterranean coast, where a blaze had swept very locally through the open vegetation on coastal dunes. The nymphs of a grasshopper common in the area, *Prionosthenus galericulatus*, are normally shades of brown, bearing a remarkable resemblance to flakes of bark. On the burned areas the author only found nymphs which were black, blending in perfectly with the charred stumps and grass-tufts of their fire-adjusted habitat. Whether these nymphs were able to change colour immediately after the conflagration or had to wait until they moulted is unknown. It is also possible that melanistic individuals were present in small numbers before the flames enabled them to become dominant after the normal forms had been culled by predators. The advantages of fire melanism are considerable, particularly in Africa, where eager throngs of insectivorous birds such as storks and herons are drawn to the fires. They stuff themselves on the easy pickings to be had as thousands of insects flee the advancing holocaust, remaining on the scene for several days to clean up any toasted corpses and capitalize on the ease of spotting previously well-camouflaged insects, now rendered easy meat in the drastically altered circumstances. Under such extreme pressure, the evolution of the ability to adapt rapidly to the fire-ravaged landscape seems inevitable.

Most grasshoppers neither feed upon nor are well camouflaged against grass. Many desert types consume a wide variety of herbs, but as they spend most of their time on the ground, they match the soil and will be discussed below. A few species are restricted to a single food-plant on which the day is spent. Under these special circumstances they may be most effectively cryptic only when on the correct food-plant and, just as important, in the correct position upon the plant. Thus *Bootettix argentatus* from the deserts of the south-western USA and Mexico feeds and lives on the creosote bush *Larrea divaricata*. It spends the day sitting among the leaves and both the colour and texture of the grasshopper are very similar to the yellowish-green resin-coated leaves of the host-plant. White marks on the grasshopper's thorax resemble the highlights formed as the intense desert sun pierces through gaps in the foliage or reflects off the shiny leaves. The adults are reluctant to forsake their bush, but can fly short distances when disturbed, disappearing instantly as soon as they again settle among the leaves. *Ligurotettix coquiletti* also lives on the creosote

Plate 76 Stick grasshoppers of the exclusively South American family Proscopiidae mimic twigs and sticks. This *Tetanorhynchus* species from the high Bolivian Andes inhabits a treeless zone where it lives among grass and herbs. It often walks across open ground where it is more or less invisible among the general detritus. All proscopiids are wingless. They can be found in deserts and rainforests as well as high mountain areas.

bush, but hunches down on the grey stems which it closely resembles.

A similar-looking but different species of creosote bush, *Larrea cuneifolia*, forms dense stands in the Monte Desert of northern Argentina, an ecological equivalent of the same habitat in North America. The Monte also has its grasshopper which feeds exclusively on creosote bush; not an acridid, as in North America, but a stick grasshopper of the family Proscopiidae. Stick grasshoppers are found in a wide variety of habitats and can often be found sitting in full view on leaves. Most kinds will freeze and sit tight if disturbed, but if really molested or actually touched they will normally throw themselves off the leaf and fall to the ground, where they hold their legs straight out to front and back and remain motionless. Lying among the jumbled mass of sticks and dead grass on the ground they are so difficult to pick out that further searching would be unlikely to pay worthwhile dividends, so the dive-for-cover last-ditch strategy is obviously a sound one. However, *Astroma riojanum*, the Monte's specialist dweller on creosote bushes, varies its defensive response according to whether the threat is to a male, a female or a nymph. It is the dominant grasshopper in the area, with up to 200 adults being present on a single large creosote bush. The females mimic the old, knotted basal sections of the stems and spend the day clustered in the lower regions of the plant in a specialized posture which makes them very awkward to discern. When disturbed they stay put, for as soon as they leave the plant their mimicry is reduced and they

Plate 77 In deserts the nymphs of many grasshoppers mimic stones, something which is also found in a few adults. This is the nymph of *Batrachornis perloides* mimicking a quartzite pebble in the Namaqualand desert in South Africa.

would be much easier to spot. The adult males are very much smaller and slimmer and sit among the thin outer twigs, which they resemble closely. They respond to disturbance by actively jumping to the ground, where their small size makes them an unrewarding target among the confusion of similar-looking detritus. The nymphs vary in colour and pattern and also jump to the ground when disturbed, where they are so hard to see that birds which have flushed them from the bushes usually give up the chase at once, and don't bother even to begin the hopeless quest for their errant quarry.

Many birds, particularly in desert areas, search for prey on the ground, or will swoop down to pick up anything which makes itself obvious by moving. Large numbers of grasshoppers live on open ground, which is often the dominant habitat in deserts, sand-dunes, heaths, the open parts of tropical dry forests and on mountain screes and lava-runs. Some grasshoppers, such as the plump *Trachypetrella* species from the deserts of Namaqualand and the Karoo in South Africa, are so like stones that an examination from very close range may be necessary before their animal status can be established. The nymphs of many of the grasshoppers from these (and other) deserts also mimic pebbles, while various species of *Lamarckiana* (Pamphaginae) resemble living nuggets hewed from the rock itself.

These 'living stones' are the sophisticates among their kind. The majority of open-ground grasshoppers are adapted for their exposed lifestyles to no greater an extent than being suitably coloured to match the substrate, their shape being the same as in most grasshoppers. The principle that simple cryptic coloration is adequate for survival is indicated by the very large number of different species and genera of grasshoppers from open ground all over the world which look superficially more or less identical. They are drably turned out in greys, browns, whites or blacks, according to the substrate, and speckled or mottled with paler colours to suit.

However, if their camouflage fails them, many of these open-ground grasshoppers can fall back on a secondary mechanism linked to their normal escape behaviour. As they catapult themselves into the air, they surprise the onlooker

Plate 78 Many grasshopper nymphs are passable mimics of dead leaves. This *Goniaea australasiae* nymph perfectly resembles a fallen *Eucalyptus* leaf in dry forest in South Australia. The fully winged adults make far less convincing leaves.

by suddenly revealing vividly hued wings, usually some shade of red, blue or yellow. The grasshopper flutters its unexpectedly colourful course across the sky for a few seconds and then plunges suddenly to earth. Here it closes its wings and instantly merges once again into the ground. Such 'flash' coloration can be seen in many grasshoppers, from the sand-dunes of southern France to the deserts of the USA or Australia and the mountains of eastern Africa, so it is presumably a highly successful survival tactic. Its success probably depends upon a number of factors. An enemy loses valuable time right at the start, being startled by the sudden unforeseen eruption of colour from a previously cryptic insect. The subsequent rapid zig-zagging flight is difficult to follow, and then the bright target colours abruptly and puzzlingly disappear as the grasshopper goes to ground. Birds are not particularly intelligent, and it is likely that having seen the showy wings they quickly form a search-image for this one bright colour, which they will then seek out – to no avail – when the grasshopper lands.

Grasshoppers are not necessarily given the chance to get off the hook using their flash display. In Kenya the author watched a bush shrike hunting grasshoppers. It sat about 6 ft (2 m) up in a bush and stared intently at a patch of bare sandy ground below. Every few minutes it would swoop down and fly back up with a grasshopper struggling in its beak. The species being caught had bright yellow flash colours on its wings, but never had the opportunity to display them, given the shrike's method of dropping in for a meal. It was cryptically coloured to match the ground but was undoubtedly betrayed by small movements, which were exactly what the shrike's piercing gaze was directed at. Kookaburras have been seen hunting in the same way in Australia, but under conditions of near darkness with the grasshoppers 6 or 7 yards (or metres) away. The ability of the bird to spot an insect in such light at such a distance does much to explain why crypsis has often reached such a peak of perfection, although in the case both of the shrike and the kookaburra it is almost certainly *movement* by their quarry which the birds' eyes are best adapted to detect.

In tropical rainforests, ground-dwelling grasshoppers are much rarer than in deserts, and those which do reside in the deep shade cast by the giant trees often resemble the dead leaves among which they live. In South America acridids of the genera *Colpolopha* and *Xyleus* are generally found on the ground among dead leaves, even in the darker parts of the forest. The nymphs are by far the best mimics of dead leaves and are more or less impossible to see until they move, while the adults are less well-concealed, being given away by their long wings which are not particularly leaf-like. This also applies to the ground-living grasshoppers in dry forests, such as Australia's eucalyptus woodlands, where nymphs of *Goniaea australasiae* are persuasive mimics of dead eucalyptus leaves, far more convincing than the fully winged adults. However, species of *Lobosceliana* in East Africa copy dead leaves perfectly both as nymphs and as adults. The author once saw a large female *Lobosceliana* rocking back and forth slightly as she layed her eggs in a sandy path. Precisely what was happening only became evident after an inspection from 1 ft (30 cm) or so away. Until then the author had been under the puzzling yet convincing impression that a dead leaf was being repeatedly tugged at by some concealed animal in order to

Plate 79 The grasshopper *Tropidostethus bicarinatus* resembles a fallen beech leaf. It lives on the ground in the magnificent forests of *Nothofagus* beeches in southern Chile.

Plate 80 Most of the African pyrgomorph grasshoppers are defended by extremely repulsive chemicals which disseminate a revolting stench around the insect. Their colours are often to a certain extent aposematic, but an extra unmistakable 'keep off' visual warning is kept in reserve for when the grasshopper is actually touched. When this happens to the large lethargic South African *Dictyophorus spumans*, it raises its forewings to reveal its brilliantly marked abdomen, while at the same time sticking its red and black legs upwards and outwards to its sides.

Plate 81 The back legs of this large *Agriacris trilineata* grasshopper from a Peruvian rainforest are amply furnished with sharp spines perfect for jabbing into an opponent's face.

drag it down into a burrow! Such plausible mimicry is not restricted to the tropics, for in the magnificent forests of *Nothofagus* beeches in southern Chile the grasshopper *Tropidostethus bicarinatus* embodies the colour and shape of a dead, curled beech leaf, and there are plenty of those on the ground where the grasshoppers lead their secretive lives.

Relatively few grasshoppers mimic green living leaves as is so typical of the katydids. When they do they can be extremely good at it. Some species of *Chorotypus* (Eumastacidae) from forests in Malaysia are among the most bizarre of all Orthoptera. The pronotum is extended upwards to form an enormous flattened shield, while the whole insect looks so flat, almost as flat as a real leaf in fact, that it seems unlikely that sufficient internal space has been reserved for the vital organs. This strange apparition occurs in brown or green forms, mimicking leaves both living and dead. Each colour-form is dominant according to the season and the resultant frequency of dead or living leaves in the environment, a brilliant example of an adaptive response to changes in the surroundings.

Even in tropical forests, many grasshoppers rely on nothing more sophisticated than merely being green to escape detection. However, significant numbers of these are extremely large, so if they are spotted they have a reasonable chance of seeing off an enemy by resorting to physical retaliation. Many kinds of large African *Kraussaria* and *Acanthacris*, or the huge South American species of *Agriacris* are armed with back legs which bristle with an imposing array of spines. On the African species these are fairly short, but very tough and sharp and suitably offset to pierce from a variety of angles, while in *Agriacris* shorter spines are mixed with long, slender, backwardly-directed, rapier-like points. The first reaction to trouble in any of these grasshoppers is to raise the back legs slightly off the leaf, 'cocking the trigger' in readiness to give a hefty kick backwards and ram the spines forcibly and painfully into an attacker's face or hands. In so doing, the grasshopper is reducing its ability to jump out of trouble, but these rather clumsy heavyweights are poor jumpers anyway and for such well-equipped pugilists counterattack is the more effective strategy.

Deterrence is the mainstay of the legions of warningly coloured grasshoppers, which deploy chemical rather than physical weapons to abort an attack. In Africa, garishly patterned members of the Pyrgomorphinae are very much a part of the everyday scene, particularly in dry areas, even in gardens in towns. Many of these are large, corpulent, clumsy, leathery-skinned insects, able to fly only poorly or not at all and mostly not particularly adept at hopping or even walking. They spend most of their time posing conspicuously on their food-plants, regularly stuffing themselves with the heavy doses of poison-loaded leaves which both nourish and protect their consumers. One of the most familiar pyrgomorphs in South Africa, a country which abounds in these gaudy creatures, is the common milkweed grasshopper *Phymateus morbillosus*. Its colour is somewhat variable, usually deep blue and red with a yellow and purple abdomen, while the crimson wings, spotted with yellow, are deliberately exposed when the insect is molested. If its ornate uniform fails to stave off an attack and the grasshopper comes in for some rough handling, it responds by spewing forth a frothy liquid from openings situated near the bases of the rear legs.

All the larger African pyrgomorphs react in a similar way, although the amount of froth produced by the other species is usually inferior to the milkweed grasshopper's copious supply. Wingless members of the clan cannot include the sudden display of brightly coloured wings as part of the 'keep off' sequence. *Dictyophorus spumans* raises its rather short forewings, the better to expose its plump abdomen, brilliantly ringed in red, yellow and black, at the same time adding emphasis to the deterrent message by sticking its black and red rear legs upwards and outwards to the sides. It would be surprising if the foam produced by these grasshoppers should ever prove ineffective in promoting a permanent stay of execution, as it really does *stink*, and even from a distance of a yard (metre) or so a nauseating stench provides a protective chemical umbrella around the grasshopper. If any animal were hungry or foolish enough to go ahead and eat one of these revolting creatures, the consequences are likely to be dire, for the display is no empty bluff. In South Africa a child died after eating just a single *Phymateus leprosus*.

The nymphs of most of these chemically protected pyrgomorphs congregate in large bands which roam around the countryside, protected by the ever-present disgusting aura which they trail around with them. The early instars are generally black, later changing to more vivid colours, often bright spots against a dark background, although never the same as in the adult. They are often noticeably successful insects. In the Sudan *Poecilocerus hieroglyphicus* is common in the semi-desert areas, despite the severe climatic conditions. It prefers to feed on a milkweed, *Calotropis procera*, whose latex forms an important attractant for the grasshoppers, although it has the opposite effect on sheep and goats, which strictly shun the plant, even when half-starved. When threatened, the grasshopper will emit a jet of whitish liquid which is extremely irritant to human eyes, as well as having a repugnant smell and taste. The fluid is ejected from a gland situated between the first and second abdominal segments and is known to be effective against lizards, toads, mantids and probably birds as well. Even in Africa, warning colours are not restricted to pyrgomorphs, for one of the most beautiful of all African grasshoppers is a *Thericles* species, a gorgeous purple creature with a bright yellow head, a distinguished member of the Eumastacidae.

Both North and South America are home to a wide range of warningly coloured grasshoppers belonging to various subfamilies of the Acrididae as well as to the Eumasticidae. Many of the most striking and beautiful are found in deserts, where the sumptuous blue and red 'painted grasshoppers', *Dactylotum* spp., may be found. A disgusting froth is also produced by some of these American species, most notably by the slow-moving, wingless lubber grasshopper *Romalea microptera*. Its first defensive measure is to hiss and foam at the mouth and thorax. This froth is produced by air mixing with the repellent chemicals. The thousands of tiny bubbles rapidly decay and burst, disseminating a protective envelope of chemical-mist around the insect, just as in the African pyrgomorphs. If this revolting stench fails to deter its opponent, the grasshopper follows up by disgorging a particularly repulsive gob of chemical-rich liquid from its mouth.

The numerous rather similar-looking species of *Chromacris* (Romaleinae) are also found in a variety of habitats, from deserts in Mexico and Argentina to

Plate 82 The *Dactylotum* species of 'painted grasshoppers' are found in the deserts of the southwestern United States and Mexico. They are usually some beautiful 'warning' shade of blue marked with orange or red.

Plate 83 The nymphs of many warningly coloured grasshoppers often wear a different 'warning uniform' from that typical of the adult. The nymphs (on left) of *Chromacris colorata* in northern Mexico may often be found aggregated on their *Solanum* food-plant in company with the differently coloured adults. The latter can fly and have bright crimson wings.

Plate 84 This large Mexican *Taeniopoda auricornis* grasshopper exposes a 'warning message' of crimson wings when molested. Note the bright scarlet mites attached to the wings. Such mites are often found riding around on the bodies or wings of grasshoppers.

rainforests in Panama and Peru and the seasonally flooded *llanos* of Venezuela. The adults are usually ostentatiously marked in green and yellow, keeping in reserve their red wings, which are prominently displayed if threatened. The nymphs are usually black with red splashes or spots and aggregate strongly, sometimes in company with adults, on the leaf of the food-plant, which is normally some kind of *Solanum* or closely related plant. The nymphs usually scatter in all directions when disturbed, but soon reassemble, although the adults fly well and can make a winged escape. The nymphs of *Taeniopoda eques* from Mexico's northern deserts are also gregarious. The adults are impressively large and splendid insects, with mustard-yellow bodies and jade-green forewings elegantly inscribed with an intricate filigree of silvery veins. If molested, this remarkable creature abruptly raises its forewings to expose the 'warning banner' of its large hind wings spectacularly patterned in red and black.

The Middle Eastern grasshopper *Pareuprepocnemis syriaca* employs a highly unusual defensive tactic. When seized, it emits a high-intensity chirping generated by its mouthparts. Lizards which were seen to grasp the grasshopper by the head (most predators go for the head initially) soon let go, allowing the grasshopper time to make good its escape. It seems that having a loud vibrating sound booming through its head tends to discourage a lizard's appetite, and the same may well apply to birds.

PHASMATODEA

Leaf insects are among the most credible of all mimics of leaves. Even the legs are flattened so that they resemble parts of a leaf, and as long as they remain still these insects are virtually impossible to discern. The same applies to the stick insects, in which physical appearance, posture, location and immobility may all be required in combination to avoid recognition. Contrary to popular misconception, these insects do not necessarily sit around on sticks, seeking to submerge their identity by posing in a sea of genuine look-alikes. In fact stick insects resident in rainforests may seldom sit in such a position, but instead repose in full view on the tops of the broad leaves typical of these forests. The success of this strategy depends on the fact that leaves and twigs rain down from the canopy on a constant basis, and the broad leaves at lower levels interrupt a lot of this detritus before it can reach the ground. Stick insects are therefore well placed to mimic these fallen twigs, often showing a remarkable ability to reduce their chances of detection by carefully placing themselves along the mid-vein on the lowest part of the leaf. This is just where most of the naturally occurring sticks will tend to end up. To be convincing, correct posture is vital, so the insect rests with its front and rear legs stretched out fore and aft and its short middle legs tucked into its sides. Many species have a particularly gnarled exterior which adds credence to the charade. Certain kinds sport outgrowths on their legs which thereby resemble peeling bark, and may even arch the abdomen upwards over the back to give the impression of a bent twig.

Many Madagascan phasmids live on trunks and branches among dense coverings of moss, which they resemble to an extraordinary degree. Some

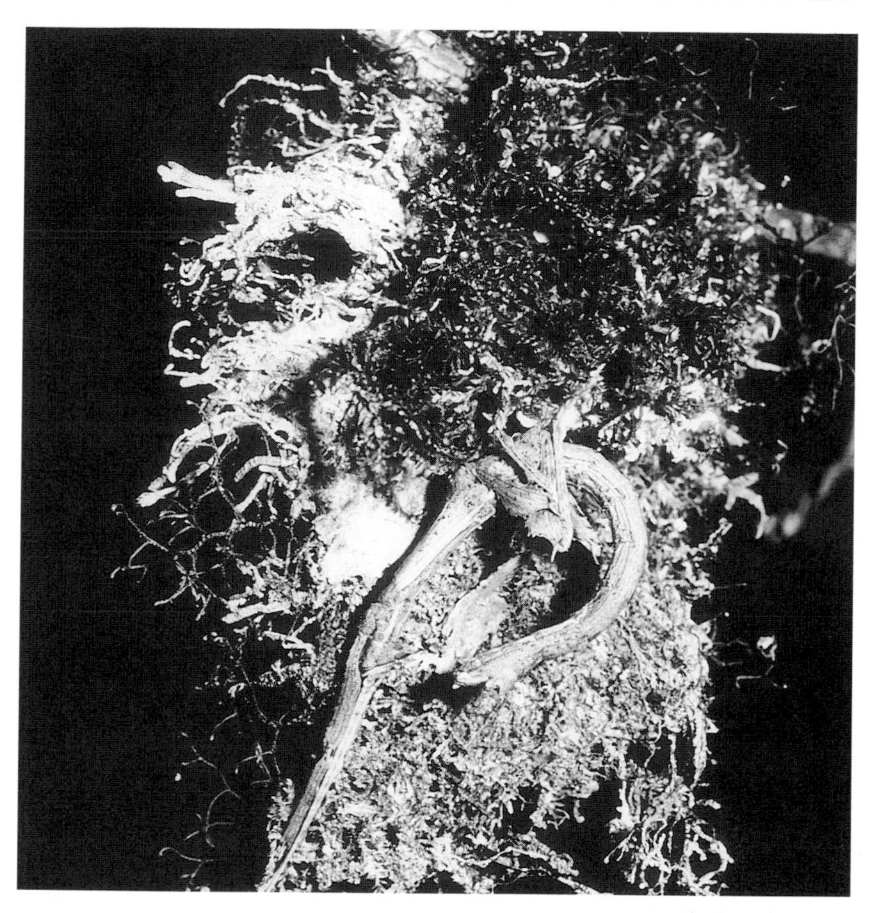

Plate 85 Location and posture can be vital for preserving a stick insect's life. This Madagascan species hangs in an inverted position beneath some mossy detritus caught on a twig. It curves its abdomen downwards almost into a circle and has extensions on the rear pairs of legs resembling flaking bark. The author found more than one species in Madagascar which adopted this habit of curving the abdomen and perching on detritus lodged in trees.

species, such as *Calvisia* from Malaysia and *Prisopus* from Central America, have flattened bodies with lichen-like outgrowths and mottled coloration, and spend the day closely pressed against a tree trunk. A few species are deep metallic blue and may be warningly coloured. Even fully winged kinds generally sit tight when disturbed, but a few will take to the air, such as a large splendid *Stratoctes* from South America which often perches on some huge leaf, unfurling bright yellow wings and sailing off into the canopy when alarmed. The majority of species, however, merely jump to the ground when touched and lie doggo until the danger passes. Others resort to defensive displays, usually involving the sudden unveiling of brightly coloured wings, mainly combinations of black and red but sometimes involving yellow, blue or orange

Plate 86 A number of stick insects spend the day flattened against the bark of trees. This is a species of *Calvisia* from rainforest in Borneo.

with black. In some species these colours are displayed continuously for a minute or more until the disturbance ceases, while in others they merely act as 'flash' colours to confuse an enemy before the insect falls to the ground and immediately conceals its wings.

The New Guinea stick insect *Eurycantha horrida* is a remarkable insect in a number of ways. Its large, stout body and stocky legs bristle with a defensive stockade of sharp prickles. When handled the male uses his well-spined back legs as a vice to impale an offending finger, at the same time curling his abdomen upwards over his back to reveal a series of transverse white stripes on the underside. The tip of the abdomen is embellished with a series of 'teeth'

Plate 87 Several stick insects 'flash' coloured wings when touched. This is a large female *Acrophylla titan* photographed in full defensive display in coastal paperbark forest in Queensland, Australia.

which resemble jaws, and as the insect rapidly flicks its copulatory organ in and out, the whole display conspires to mimic an open-mouthed, flicking-tongued serpent ready to strike. It backs up the whole display by releasing a brown liquid with an offensive odour. A number of large stick insects cock their tails over their backs when molested in an action which is thought to mimic a scorpion about to strike, while nymphs of one or two species mimic ants. The large southern walking stick *Anisomorpha buprestoides* of the southeast USA goes one step further in its application of chemical deterrence and can actually direct a defensive spray backwards with great accuracy over a distance of some 16 in (40 cm). This is effective against birds, mice and mantids but less so against a mouse possum *Marmosa demararae* which holds the insect out of harm's way in one paw until its unpleasant spray is completely discharged, when it eats the insect.

Chapter 6
Where They Live

The orthopteroid insects enjoy a wide distribution around the world, although the richest diversity is found in the tropics, and they are absent from regions of permanent cold. They can be found in virtually every kind of habitat, from the wave-swept seashore to the humid gloom of tropical forests, from parched lowland deserts shimmering in a haze of heat to grim mountainsides chilled by frequent blizzards of sleet driven before an icy wind, to altitudes of 14,000 feet (4300 m) or more. They may be the most important consumers of the grasses in coastal salt-marshes and terrible defoliators of trees in tropical dry forests. A few kinds manage to live in ponds and lakes, but the open sea has kept them at bay.

Despite their obvious success, no types are found in every kind of habitat, and ecological factors strongly influence their overall spread. Certain pest cockroaches have attained a cosmopolitan distribution with man's help, but in cooler climates they are restricted to artificial habitats heated by man, replacing the natural warmth of their former tropical home. Heat is but one factor which restricts an insect's ability to survive in an area. The amount of rainfall is also important. An insect adapted for living in desert conditions would not thrive if transferred to a rainforest, whose own denizens would not last long in the parched, sun-drenched landscapes of a desert. There are exceptions and some insects are remarkably adept at doing well in a variety of environments. The elegant grasshopper *Zonocerus elegans* in South Africa seems to be equally at home in deserts, savannahs, dry forest, subtropical rainforest, cultivated croplands and gardens. This species has a catholic diet, so it is not forced to live only in areas suitable for the growth of a certain species of plant, as happens to grasshoppers with restricted dietary preferences.

Even within a preferred type of habitat, for example grassland, confined to a relatively small area such as southern England, micro-climatic factors may intervene which restrict the distribution of a species to certain especially favourable localities. Thus the stripe-winged and rufous grasshoppers *Stenobothrus lineatus* and *Gomphocerippus rufus* are thought to be restricted to southern England on climatic grounds – further north the summers are too cool and wet and the winters too long. However, although grasslands of one type or another, all man-made or man-modified, occur throughout southern England, these two grasshoppers are restricted to areas of fairly short grasses growing on shallow soils overtopping chalk or limestone rock. Even within this limitation it is often only steep, south-facing slopes which receive and trap the maximum amount of heat that are colonized. And even where these are available and everything looks perfect for the grasshoppers, they may be puzzlingly absent, or only the stripe-winged may be present, as it seems slightly more tolerant of less-than-perfect conditions than the rufous. Yet the similar-looking field grasshopper *Chorthippus brunneus* lacks these ecological shackles and can be found not only in the type of short grassland preferred by the above two species, but also on heaths and moors on acidic soils, grassy rides in woods, damp pastures and even on grassy roadsides in towns. The large marsh grasshopper *Stethophyma grossum* is even more ecologically restricted. Not only is it only found on acid bogs, it is confined to the very wettest areas where there is free-standing water.

Sometimes small but significant ecological differences can be used as an aid to distinguishing between species which differ only marginally from one another in external details. In the United States, the three crickets *Nemobius fasciatus*, *N. allardi* and *N. tinnulus* were for many years all regarded as being members of a single species, *N. fasciatus*. However, detailed observations have established that in reality there are three completely distinct species which are separated by details of song and ecological preference, as well as by small but constant anatomical differences. *N. fasciatus* lives in rather wet situations, often among rank marsh vegetation in poorly drained pastures or along riverbanks. It is widespread over virtually the whole of the eastern part of North America from southern Canada to Florida. Its song is a series of chirps with a distinctive buzz to them, making it very different from those of the other two species. *N. allardi* prefers its home to be well-drained, with a rather open cover of grass and weeds, such as can be found on lawns, roadsides and sloping fields. It is restricted to the northeastern United States and sings with a clear musical trill. *N. tinnulus* confines itself to rather dry woodlands, often of mixed oak and hickory or oak and pine in the central part of the eastern United States, where it usually scuttles around beneath the leaf-litter. Its song is also a musical trill, but slower than in the previous species so that the pulses of sound can often be heard as separate units. The combination of habitat preference, song and disparities in close-up courtship procedures and sexual pheromones serve well to keep these three similar-looking species more or less isolated from one another, even when their ranges overlap and they live as close neighbours. Colonies may in some locations be separated only by a few yards, and yet there is no intermixing, as this would mean crossing over to an unsuitable habitat, while the difference in the songs ensures that there is no temptation to do this.

Some orthopteroids carry this ecological narrowness to extremes. In the Hawaiian Islands, there are examples of habitats moulded in the crucible of cataclysmic volcanic activity, when frequent incandescent eruptions give rise to barren moonscapes of fresh lava flows. It takes six months or more before the first macroscopic plants manage to gain a foothold on this sterile, sun-baked terrain, which seems innately inimical to any form of life. Yet even before then an amazingly specialized pioneer, the cricket *Caconemobius fori*, will have colonized this new-born landscape, scratching a living by feeding on the debris delivered by the wind from nearby better-vegetated islands. So far as is known, this remarkable insect is restricted to these barren lava flows near Kilauea Volcano, although even here it probably does not have everything its own way, for a native *Lycosa* wolf spider shares its habitat and, needing live prey, presumably relies on the cricket for most of its food.

A few species of cockroaches seem to be confined to living as commensals in birds' nests. *Euthlastoblatta facies* from Puerto Rico establishes large populations in the twiggy nests of the grey kingbird, *Tyrannus dominicensis*. The nests of *Foudia* species in Madagascar seem always to harbour large numbers of cockroaches, which are believed to live nowhere else. This kind of micro-habitat probably merits closer examination, for of ten species of cockroach found in the pendulous nests of an oriole in Brazil, two were new to science and one of these merited the erection of a new genus. All these cockroaches probably glean a living as scavengers in the nest, so presumably they are beneficial to the birds in helping to keep the nest clean and in a sanitary condition.

Birds' nests are one thing; ants' nests are something quite different. Birds don't generally poke around in their own nests looking for a bite to eat, so as long as a stowaway cockroach keeps out of sight, it should be reasonably safe. Ants' nests, on the other hand, are teeming warrens filled with hordes of ferocious guardians, ever alert and ready to pounce upon any trespasser who cannot gain *bona fide* admission by possessing the correct pass, in the form of a desirable scent or coded set of reactions. Yet despite the apparent hostility of a fortress-city in which every resident is a potential hazard, the relative stability and continuity of this food-rich habitat from season to season makes it a sought-after residence for those cunning or tough enough to survive its dangers. Large numbers of insects make the grade and live as obligatory guests within ants' nests, and are usually found nowhere else. Some of the most interesting of these myrmecophilous insects are the diminutive crickets of the genus *Myrmecophila*. *M. manni* occurs in rather dry areas from southern Idaho to northern Mexico. This cricket has been found in association with at least 13 different species of ant, but in southeastern Washington it seems to live mainly in the nests of the western thatching ant *Formica obscuripes*. The crickets seem remarkably proficient at existing side by side with the ants, for up to 50 can be found within a single nest, while a third of the nests in a given area will usually harbour this uninvited guest. Despite constant meetings between ant and cricket within the nests, the ants never seem to get used to the intruders, and always behave aggressively towards them by pressing home an attack. The crickets are constantly vigilant and take advantage of their ability to retreat at a fast sprint to outpace the ant and escape. On other occasions, however, the circumstances of meeting may enable the cricket to steal a march on its host and persuade it, by going through the correct sequence of motions, to come up with a meal in the form of regurgitated droplets of food. The female crickets lay their eggs in the soil where they are safe from discovery by the ants. The new generation appears in a mass-hatching which is synchronized in all the colonies within an area. The crickets spend a great deal of time following the ants' foraging trails at night and probably use these to move home by switching across to the trail leading to another nest when two trails overlap.

The nests of birds and ants may themselves be restricted to a certain kind of habitat, so even living in a highly specialized micro-habitat may still involve limitations on overall distribution. It is now worth looking more closely at three important habitats for orthopteroid insects, all of them very familiar to the author, namely rainforests, deserts and mountains.

RAINFORESTS

As the tropical rainforests hold the major proportion of the total number of animal species, it is not surprising that they are fascinating places for the entomologist. Every niche within a forest is exploited by some kind of orthopteroid, from cockroaches living in hollow trees, leaf-litter or birds' nests to grasshoppers ravaging the plant-life near the ground, katydids hiding away during the day among the rosettes of epiphytic bromeliads high up in the canopy or camouflaged mantids prowling on a tree-trunk. However, not all rainforests yield some exciting new marvel at every turn of the path. The author had to spend an inordinate amount of time during six weeks in the Malaysian forests to turn up relatively few species of grasshopper and katydids, and not a single mantis. Species-diversity in these forests is high, but the actual density of individuals is often disappointingly low. Orthopteroids seem to feature far more abundantly in African rainforests, while certain forests in South America are so staggeringly rich that it is difficult to decide which particular specimen to photograph out of any number of brilliantly enticing species, so great is the variety over even a small area.

Just such an entomological treasurehouse can be found in the national park at Tingo Maria in Peru, on the eastern flank of the Andes, where the lowland Amazonian forest starts to climb up the lower slopes of the mountains. Anyone who has only seen mantids or grasshoppers in Europe or the northern United States can have little grasp of what it is like to stroll down a path in this fabulous forest, of the excitement and stunned admiration inspired by the new wonders encountered at almost every step. Many of the more resplendent grasshoppers almost beg to be appreciated as they loll around conspicuously on every other leaf, although the more perfectly cryptic of the mantids and katydids have to be worked for and demand some careful searching. Even so, some of these are so common at Tingo Maria that they are easy to spot once the enthusiast has formed the correct search-image. 'Getting one's eye in' is therefore of key importance. The reader can be given no better idea of what a tropical forest can offer to the naturalist than by accompanying the author on a walk of just a few hundred yards at Tango Maria.

A track wide enough for vehicles runs below the base of a damp cliff covered in forest and clothed with a riotous growth of ferns and mosses. On the other side a narrow piece of forest separates the path from a river. Rotting logs or newly-fallen trees are scattered around under the trees or by the path and must be carefully searched. The whole area is filled with the sound of insects and birds, while sumptuous butterflies frequently flash past. Around the edges of a man-made clearing near the park entrance stands a number of trees, their trunks a confusing mosaic of lichens in greys, whites, greens, browns, blacks and pinks, mixed with the fresher green of mosses. Such heavily patterned trunks are always worth very detailed investigation anywhere in the tropics, for the communities of animals which live in this relatively austere and exposed micro-habitat are remarkably large and diverse. Here at Tingo Maria, several small cryptic *Liturgusa* species mantids soon turn up, as they do virtually anywhere in Central or South America on bark in tropical forests. Sharing their home are some small, flattened, unidentifiable cricket nymphs, blending imperceptibly into the speckled bark. The real prizes after a patient search are several large *Acanthodis aquilina* katydids, low-bodied grey and silver shapes, looking like nothing more significant than humped imperfections on the bark and invisible to all but the most practised of eyes. One trunk is clothed

Plate 88 The majority of the world's insects live in tropical rainforests, such as this particularly rich area near Tingo Maria in Peru. The walk described by the author is along the base of the slope on the right of the picture. This particular stretch of forest probably holds the grestest variety of orthopteroids to be found in any single restricted locality in the world.

in an exuberant coat of mosses. Hunched down among these is a deeper-bodied katydid *Championica peruviana*, its colour and pattern exactly matching its surroundings, while the serrated-edged, upturned pronotum resembles a jutting tuft of mosses.

A 'dead leaf' suddenly leaps out from beneath a descending boot. A *Xyleus* grasshopper nymph has avoided death by its quick reactions, while a small distance away sits a mating pair, less like a dead leaf than the nymph, but still noticed purely by chance as they rest quietly on a shrivelled leaf. Disturbed by the author's tread, a shiny brown and cream *Proscratea peruana* cockroach skates rapidly under cover beneath a fallen branch. The ground hosts its share of species such as this, but it is the vegetation up to around a yard (metre) in height, as well as the overhanging branches, which now require a really close investigation. Many rainforest grasshoppers naturally live in the more open glades created by the toppling of some giant tree, which may enlarge the gap considerably by felling several close neighbours in its tempestuous descent. This relieves the pervading gloom of the undisturbed forest and allows a vigorous growth of plants to burst forth in the flood of sunlight. Such favourable spots are generally quite rare and widely scattered under natural conditions, but where man has intervened they may occur on a much larger scale. The track at Tingo Mario in effect creates a 'linear treefall' in the forest. It satisfies the criteria needed to favour a very high diversity of species – not so narrow that the sun is still faint and few plants can flourish, nor so wide that it becomes too hot to be endured during the hottest part of the day.

Many of the understorey plants have much larger leaves than would be found in a similar situation in temperate forests, so grasshoppers using them as perches may be very easy to pick out. At the opposite extreme, a long, slim, unnamed katydid is quite common at Tingo Maria, lying along the mid-veins of the large aroid leaves during the day, especially under the dripping wet, shady cliff. In its specialized posture and carefully selected position it takes a very skilled eye to spot it. An *Oxyprora* katydid adopts a similar pose, but it sits up higher on its legs and poses fewer problems in detection.

As we enter a well-lit stretch, the grasshoppers start to appear in exciting variety and dazzling colours. Several slender *Stenopla boliviana*, elegant in polished green and black, are perched on the long, broad leaf of a *Heliconia*. On the same plant, but almost out of reach, sits a mating pair of *Rhopsotettix consummatus*, gaudily marked in a warning uniform of black and yellow polka-

Plate 89 A pair of warningly coloured *Rhopsotettix consummatus* grasshoppers pose on a leaf in rainforest at Tingo Maria in Peru.

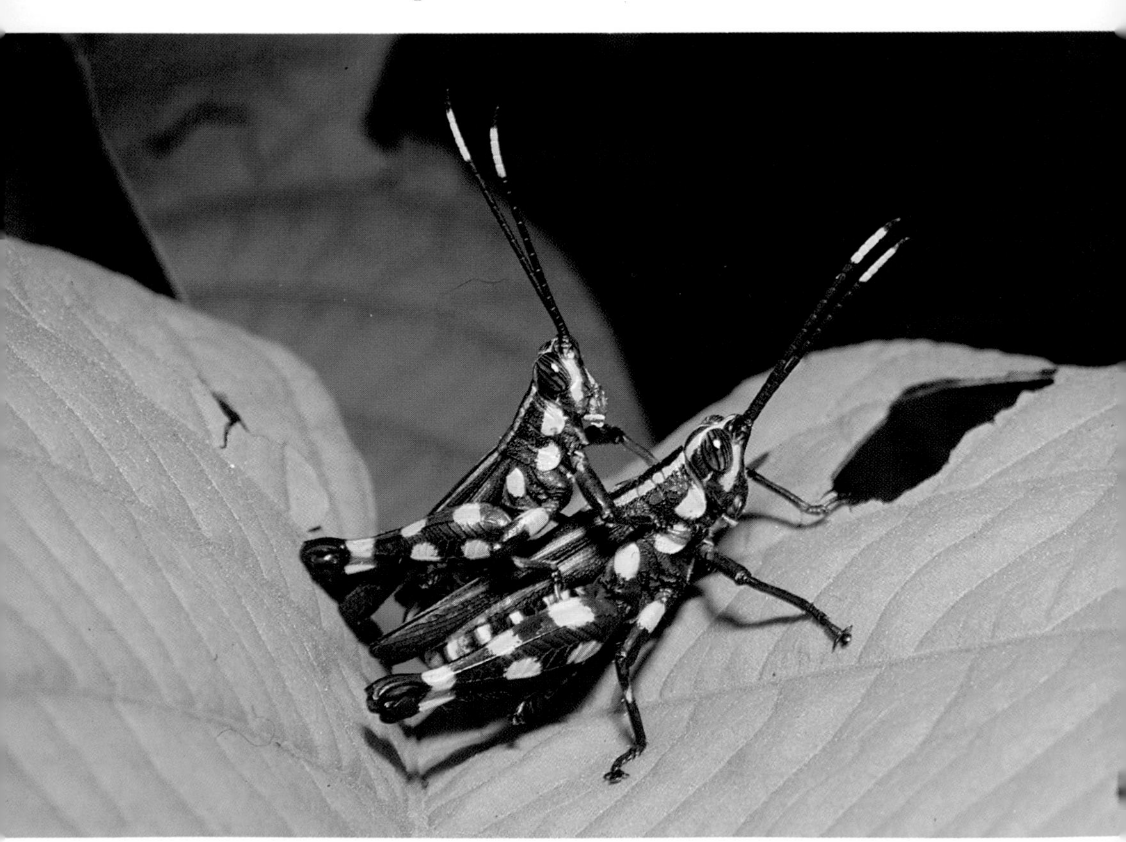

Plate 90 *Tetrataenia surinama* is but one of a medley of brilliantly attractive grasshoppers which can be easily found in a short walk through the rainforest near Tingo Maria in Peru.

dots. Just a turn of the head reveals a male *Tetrataenia surinama*, a living bauble with his brilliant green back contrasting with the red, black and yellow of his 'golfing pants' legs. Sharing his leaf is a mating pair of a daintily pretty eumastacid, *Eumastax pictipes*, with pale blue and yellow sides and green forewings. A little way away from the happy couple sits another mating pair; at first glance they seem to belong to the same species, but a closer look reveals black wing-cases and less blue on the sides. It looks different and proves to be *Eumastax jagoi*. Just a few steps further on, the leaves of a *Solanum* bush have been torn to shreds. The culprits, *Chromacris mites*, are plain to see, for they are busy adding to the damage and sit there munching away, flagrantly parading their green and yellow warning colours. The leaf nearby is not in much better shape, being attacked by a small undescribed species of *Eusitalces* with dark grey back, lime-green legs and startling pinkish-orange eyes. This plant seems to be favoured by more than one type of grasshopper, for its leaves are also savoured by a coffee-coloured, rough-textured *Lamiacris nigroguttata* and a gangly-limbed brown *Pseudomastacops* eumastacid with white-blotched sides. A pair of rotund harlequin-patterned black and yellow leaf-beetles attract the eye to another leaf where sit, previously unnoticed, two different species of small *Psiloscirtus* grasshoppers, both of which are probably undescribed. One of the species is mating, a rather drab insect in this kind of company, pale brownish with green legs; the other is prettier, with a high-gloss, almost glass-like dark green body, yellow front legs and salmon-pink eyes.

A little further on, the trees close in somewhat on either side to create moister, more shady conditions. On the lower leaves of a tree overhanging the path sits a mating pair of one of this forest's biggest grasshoppers, *Prionolopha serrata*, pale green with a front end shaped like the prow of a boat and spiked back legs. A 'twig' resting on a nearby leaf starts to walk off as a female *Proscopia* stick grasshopper prepares to dive for cover on the ground. For the next hundred paces it is essential to examine all the accessible leaves, for this is where some of the nocturnal katydids spend the day in a cryptic pose. At face height perches an *Orophus* katydid, a beautiful mimic of a green leaf; it is still on the same leaf four days later. A careful inspection turns up two more katydid mimics of green leaves sitting lower down, *Stilpnochlora incisa* and *Cnemidophyllum stridulans*. Both these katydids are reasonably convincing mimics, but being quite large, they are not really difficult to spot with an experienced eye.

Considerably greater expertise is required to detect the male *Acanthops falcata* mantis looking for all the world like a dead leaf which has drifted gently down from the canopy and come to rest on a broad *Heliconia* leaf. This painstaking searching also pays off with a lovely cockroach, *Paratropes lycoides*, sitting in full view on a leaf near the ground and at first mistaken for a fallen fruit. In close proximity sits a grasshopper, *Aptoceras carbonelli*, whose green and grey speckling and scabrous texture give it the appearance of a small lichen-covered twig which has dropped on to a leaf. In fact this grasshopper is far less obvious than the slim, dark green male *Stratoctes forcipatus* stick insect sitting draped along the middle of a leaf. He has orange markings near his wing-bases, and may be warningly coloured, which could explain why his twig-mimicry is inferior to that of many other insects in this forest. This might also apply to the large 8 in (20 cm) long unidentified *Stratoctes* which is not particularly cryptic, and the black and white banding on its legs, body and antennae and brilliant yellow wings suddenly displayed as it flies off point towards possible chemical defences. Two more of these gentle giants rustle mysteriously into the air in the next few minutes, followed by the discovery of a huge green female *Agriacris trilineata* grasshopper wrecking a leaf with her powerful jaws, her enormous back legs bristling with a fearsome arrangement of spines. She is in a more open area, where a large, pale green female *Macromantis* mantis has just finished attaching her papery ootheca to a twig.

A male *Adrolampis rubrovittata* grasshopper with his green sides and chocolate back bordered with strawberry-red stripes vibrates his back legs nervously as he basks on a fallen tree. And on a patch of evil-smelling ground gleam a scatter of *Paramastax nigra* eumastacids radiant in green, blue and yellow, accompanied by just one lone orange and blue *Cercoceracris viridicollis*, perhaps the most beautiful grasshopper of this forest.

This enjoyable stroll has taken us perhaps 400 yards (360 m) and two hours, accompanied by a great deal of meticulous observation. We have encountered a total of 33 species of orthopteroid insects. Compare this with the total of 29 species which are native to the whole of the British Isles. Also bear in mind that several of the better-camouflaged species will have been missed altogether,

Plate 91 This large *Stratoctes* species stick insect readily takes to the air when disturbed, exhibiting bright yellow wings. It is often seen in the rainforest near Tingo Maria.

and more intensive collecting methods would doubtless have revealed many additional species, especially crickets and cockroaches hiding away during the day under logs or debris. Interestingly enough, virtually all the grasshoppers mentioned above are flightless, a situation which is also typical of the grasshoppers resident on mountains. One could hardly envisage two habitats as disparate as a tropical lowland rainforest and an open mountainside. However, they both have one factor in common – a hindrance to easy flying. The forest is a closed environment, difficult to penetrate by any but good fliers – and few grasshoppers are that. Mountainsides are prone to high winds and unpredictable storms, two good reasons for not trusting to wings for getting around.

DESERTS

For anyone who has not experienced walking through deserts, it is difficult to imagine just how rich they can be in plant and animal life. Many American readers will know exactly what is meant by this, but British naturalists frequently struggle to visualize deserts as offering up little but sand and rock. It is therefore somewhat ironic to learn that a relatively small area of parched wilderness in Arizona can support a greater diversity of orthopteroids than the entire damp and verdant British Isles.

However, not all deserts are equally productive, even when they share similar aspects of topography and ecology. Detailed comparisons have been

Plate 92 The grasshopper *Phaedrotettix valgus* was abundant in desert scrub in northern Mexico when the author stopped for a brief search of the area. Several pairs of this wingless species were mating, showing that in this species the males are only marginally smaller than the females. These North American deserts are much richer in grasshoppers than their South American counterparts.

made between areas of the Sonoran Desert in Arizona dominated by creosote bushes (*Larrea*) and ecologically equivalent areas in the Monte region of Argentina. The Sonoran Desert came out way on top with 27 species of grasshoppers in the study area, compared with only nine species in the Monte. The range of plants in the two deserts is very similar, with several of the dominant genera being shared. The reasons for the inferiority of the Argentinian totals are unclear, but probably have to do with the lesser age of the deserts in that region, their more variable climate over long periods and the scarcity of orthopteroids in closely adjacent mountain habitats which would reduce the chances of recruitment of new species into the Monte. This is in accord with the author's own experiences of looking for grasshoppers in the drier parts of South America, which are quite unrewarding places for insects compared with the equivalent areas in North America.

A brief roadside stop which the author made in the Chihuahuan Desert in Nuevo Leon State in northern Mexico might help to illustrate the point. The habitat consisted of low thorny scrub alongside areas of creosote bush, grading into the stony bank of a dried-up river. There had been a prolonged drought and the plants were all shrivelled and dusty, yet the ground beneath the scrub was teeming with *Phaedrotettix valgus* grasshoppers, which catapulted off in all directions as one walked through. This is a fairly small but handsome wingless species with a prominent white face, white sides neatly embellished with a broad brown stripe, and vivid green legs. On the bushes themselves sat one or two huge majestic *Taeniopoda eques* which flashed their bright red wings in a warning gesture if the author came too close. The resinous leaves of the creosote bushes supported masses of *Bootettix argentatus*, which has already been mentioned in previous chapters, and another smaller bush with similarly resin-coated leaves played host to *Hesperotettix viridis*, an attractive olive-green grasshopper with bright red markings on its legs. A *Solanum* plant above the riverbed was being rapidly shredded by colourful groups of *Chromacris colorata*, the adults decked out in a 'warning' livery of greyish-green with contrasting splashes of yellow, the crowded nymphs conspicuously florid in glossy black and red. Among some dead sticks on the ground lurked several *Mermiria bivittata*, elongated slim beige and grey grasshoppers with very long back legs and surprising black and white spotted eyes. On the ground lazed a single *Dactylotum* painted grasshopper, resplendent in blue, yellow and red, easy to see, and many dirt-coloured *Cibolacris parviceps*, giving themselves away only when they jumped. On bare stony slopes above the river were *Pedioscirtetes maculipennis* with an unusual brown and white chequered pattern, long slender *Mermiria texana*, very elegant with its black and cream longitudinal bands, and *Platylactista azteca*, dull grey, like the rock itself. However, the prize discovery here was a species of *Phrynotettix* which exactly mimicked a pale limestone pebble.

Mantids seem much scarcer in these North American deserts than in their

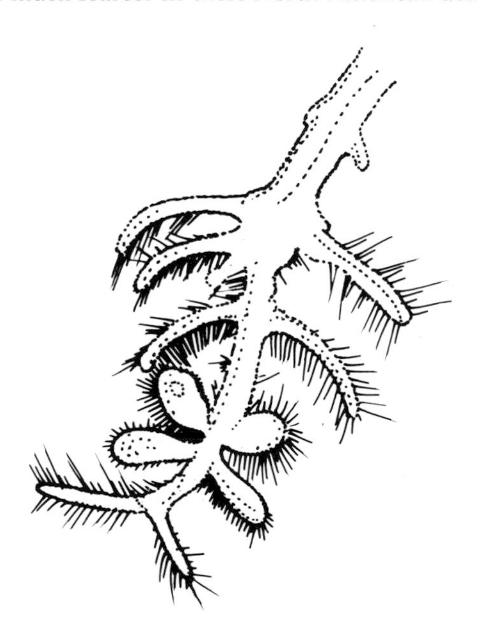

Fig.11 The hind foot of a *Comicus* Namib Desert dune cricket showing the remarkable adaptations for walking on unstable sand.

African counterparts, although the variety of Orthoptera which can be casually encountered in the two areas is similar. The deserts in South America are noticeably poor in all orthopteroid insects, even the Atacama which is undoubtedly the oldest among them. The Namib Desert in Namibia is extremely ancient and although the large areas of bare rolling sand-dunes are unsuitable for colonization by many orthopteroids, the *Comicus* Namib Desert dune crickets make this inhospitable landscape their special home. They hide in burrows in the sand during the day and emerge at night to feed. They move confidently over the loose, unstable sand, assisted by their remarkably modified tarsi, especially those on the hind legs. These 'snowshoe feet' provide traction on the sand through being split into a series of flattened lobes densely fringed by hairs. This is one of the most extreme examples of an insect adapting to fit a specific habitat.

Plate 93 *Punacris peruviana* is a member of the subfamily Tristirinae found at very high altitudes in the Peruvian Andes. Like most mountain grasshoppers it is wingless.

Plate 94 'Little friars' covering the paramo at 14,000 ft (4300 m) in the Sierra Nevada de Merida in Venezuela. Two fascinating endemic grasshoppers take shelter in the felty rosettes of these amazing plants.

MOUNTAINS

Mantids and stick insects are rare in the higher zones on mountains, where the treeless, windswept slopes are home to a range of specialized grasshoppers and a few ground-dwelling katydids and crickets. Some mountains support far greater numbers than others. A walk across the upper slopes of the Snowy Mountains in Australia is unlikely to yield much, although you are liable to see the strange, brown, wingless pyrgomorph *Monistria concinna* with its dense peppering of golden spots. It is oddly similar to *Punacris peruviana* (Tristirinae), which is found in short wiry turf in the dry Peruvian Andes at over 14,000 feet (4300 m), one of the few grasshoppers living in the inhospitable terrain of the high puna.

One of the most spectacular mountain vistas in the world is found in the Sierra Nevada de Merida in Venezuela. Here, at a height of over 13,000 feet (4000 m), is a dramatic and wonderful region known as the paramo. Assailed

Plate 95 A mating pair of *Meridacris mucujibensis* grasshoppers sheltering on an *Espeletia* rosette in the Sierra Nevada de Merida at 14,000 ft (4300 m). Without the protection afforded by these plants these grasshoppers could probably not survive in such an inhospitable area.

at any time of the year by frequent freezing storms of sleet or snow, the dun flanks of the mountains are brightened by a shining silver carpet, a living tapestry comprising millions of frosty-white rosettes of plants known locally as 'little friars'. These awe-inspiring plants are placed in the genus *Espeletia* and belong to the same family as the familiar daisies and thistles. The leaves are covered in a dense felt of silvery hairs which act as a thermal screen, a natural overcoat, designed to protect the internal tissues from the harsh year-round treatment meted out by the unpredictable climate.

Small wonder, then, that insects are noticeably absent as the walker strolls among the shining constellations of cosily muffled plants. Yet these are the cornerstones of survival for two of this region's small band of insects. For down at the base of the statuesque rosettes, snug among the soft fuzzy pile, just where the silver leaves are designed to reflect any available warmth for the plant's own use, live two species of remarkable grasshoppers which can be found nowhere else, *Meridacris mucujibensis* and *M. meridensis*. They are both rather variable in colour, and may often be eye-catchingly beautiful in shades of purple, deep blue or deep green, often set off by discreet yellow markings. During sunny intervals these hardy grasshoppers forsake their life-saving survival-shelters among the rosettes to feed and bask near the ground, but they quickly head for home when the weather turns nasty. The extra warmth and protection provided by their adopted lodgings are probably vital to the grasshoppers, enabling them to fit in vital activities such as mating, which usually takes place on the plants, during even brief warm spells. It is almost worth making a special visit to these high fastnesses just to see these incredible plants and their miniature tenants, although in general the South American mountains are poor in insects and the average person would be lucky to see more than the odd grasshopper in a day's walking.

This would certainly not be the case on most of the European mountains, such as the Alps. A stroll across a flowery Swiss mountainside in August would be accompanied by the constant serenading of countless grasshoppers, which

Plate 96 A mountain valley in Andorra in the Pyrenees in Europe. The hay-meadows in the valley-bottom and the mountain slopes beyond all contain a rich variety of grasshoppers and katydids.

Plate 97 The grasshopper *Podisma pedestris* lives in the Pyrenees in Andorra as well as further east in the Alps. Like most mountain grasshoppers it is wingless. Isolated populations of this species exhibit a genetic incompatibility which suggests that their evolution into a number of distinct species is well under way.

would establish their corporeal as well as acoustic presence by leaping madly in all directions at every step. Admittedly the colours of these energetic singers are not exactly brilliant, but a number could be considered rather pretty, especially by our modest European standards. Some are high altitude specialists, such as *Arcyptera fusca*, a handsome bottle-green species with blackish forewings, black and white banded femora and bright red tibiae; the shiny lime-green *Euthystira brachyptera; Stauroderus scalarus* with its rather muted grey-green body and dark wings; the dainty grey and black *Podisma pedestris* and the charming little vivid green *Miramella alpina*. Another dozen or so species of grasshoppers will often be present, along with five or six species of katydids – altogether a rich haul for a mountain area, particularly when compared with South America. The most abundant of the katydids will probably be the wart-biter *Decticus verrucivorus*, a heavy-bodied green insect patterned in black spots. Alpine pastures are often crawling with them, yet in the British Isles, which lie right on the northern edge of its range, this insect is exceedingly rare, being restricted to one or two tiny areas of grassland where the micro-habitat is sufficiently favourable to allow its continued and apparently rather precarious existence.

The Australian, Peruvian and Venezuelan mountain grasshoppers mentioned above are all flightless, as are many of the European species. The flightless condition is characteristic of insects which are adapted for living on mountains. Around half of the insects inhabiting the higher pastures of the

Himalayas are completely wingless, while in the Orthoptera the proportion reaches 95%. The lack of wings is presumably partly in response to the short season available for development from nymph to adult, for the extra resources needed for the production of full-length wings are simply not available in the often brief summer season. Wings may also be less useful on exposed mountain slopes anyway, where frequent strong cold winds may sweep the insect away, even if it has the energy to fly once it has left the warmer micro-climate near the ground. However, lack of wings also reduces the ability to colonize new areas, as well as limiting the exchange of genetic material between individuals in separate colonies. As suitable mountain habitats tend to be fragmented anyway, this could be a factor leading to speciation in mountain grasshoppers. Individuals from different genetic populations of *Podisma pedestris* in the Alps give rise to genetically inferior offspring when they are (artificially) mated. This seems to indicate genetic drift, leading inexorably to different colonies of this species eventually becoming genetically incompatible, new species in fact, a situation which apparently has already arisen among the *Erebia* butterflies in the Alps.

Grasshoppers which are found on mountains do not necessarily belong there, as species from lusher habitats lower down may land up on some snowy peak during a mass migration. The changeable mountain weather dooms the migrating swarm to extinction when warm sun changes within minutes to bitterly cold winds, forcing the flyers to land, even if there is nothing more friendly than snow to receive them. Once down they will never to able to leave, for the low temperatures on the snow refrigerate the insects, even if the sun again shines. The American migratory grasshopper *Melanoplus sanguinipes* is often the species involved. On Grasshopper Glacier in Montana thousands of grasshoppers are entombed in the ice, which relinquishes a number of its captives during the summer melt. The stock of frozen cadavers is frequently augmented by new arrivals, although some of the corpses are thought to be 300–400 years old.

Chapter 7
Enemies

VERTEBRATE ENEMIES

With the exception of the chemically protected species (and even some of these succumb to certain predators), the varied members of the orthopteroid group probably qualify as some of the most tasty, nourishing and generally desirable food items for many birds, lizards and mammals. Even kinds which could be thought to be relatively secure, by concealing themselves in subterranean burrows, may bear the stamp of culinary approval to the point that a pair of European hoopoes was once noted feeding their young on an exclusive diet of mole-crickets. Locust swarms are often accompanied by flocks of birds which gorge themselves until they are too heavy to take to the air. A stork faced with such a plenteous harvest can pack away 4 lb (1.8 kg) of locusts in a day's feasting.

Grasshoppers fleeing from the trampling hooves of large grazing mammals probably constitute the major part of the diet of birds such as the cattle ibis, while the bustard carefully quartering the broad expanses of Africa's savannahs will probably end up with more grasshoppers in its crop than other kinds of prey. This may be partly due to grasshoppers being the insects most frequently encountered. But they also have the advantage, from a consumer's viewpoint, of being soft-bodied, easier to handle than, say, a slippery-backed beetle, very nourishing, and, with a few spiny-legged exceptions, incapable of fighting back. Even retaliatory abilities may gain little profit against certain particularly formidable predators such as the African ground hornbill *Bucorvus leadbeateri*. Small parties of this large, stocky, black and red bird pace slowly across the ground in line-abreast formation, poking into every nook or cranny for a possible meal. If it encounters something capable of putting up a fight, such as a large *Tenodera* mantis, the hornbill takes full advantage of its most useful asset, its large curved beak. This handy piece of hardware can deftly manipulate even the largest of kicking, struggling mantids. The beak's polished horny surface is impervious to the victim's scrabbling attempts at inflicting wounds with its spiny front legs, which are kept well away from the bird's vulnerable eyes by the beak's great length.

The types of enemies which an insect has to face will vary according to its habitat. In deserts a great variety of birds will comb through everything thoroughly, from the ground to the herbs, shrubs and trees. The majority of desert lizards are terrestrial and will snap up a passing grasshopper or, perhaps, alerted by the insect's movements, run a short distance to take it. Desert foxes and many mongooses will have a go at eating just about anything, especially during hard times when even a grasshopper is better than nothing. At night, deserts harbour a rich fauna of small rodents which will use their acute sense of smell to track down grasshoppers, crickets or stick insects. In forests birds will still be the chief foe, while lizards such as chameleons use their long, extensible tongues to bridge the gap between themselves and their prey. A chameleon can sit and look straight at a perched katydid without noticing it, but as soon as the insect moves the chameleon's eye-turrets swivel forwards and the tongue zaps the prey in an instant.

An additional and very important hazard in many forests is posed by monkeys, which may occur in remarkably high densities. Being primates, monkeys are blessed with a high degree of intelligence. Combine this with excellent vision and highly developed manipulative skills and you have a very efficient hunter. They thus pose a major hazard to any edible invertebrates, especially as a large troupe of monkeys will scour the area of forest within its territory day after day, rifling through every feature for signs of edible material. Grasshoppers, crickets, katydids, cockroaches, mantids and stick insects all seem to be greatly relished by monkeys, some of which will even chew their way through bunches of grasshopper nymphs usually considered to be warningly coloured. Squirrel monkeys will sit on a branch and methodically ransack a large epiphytic bromeliad, pulling out fistsful of dead leaves and twigs which have fallen into the rosettes and painstakingly examining everything for signs of life. Leaves, both living and dead, will be turned over and over and inspected with intense concentration, hinting that monkeys know all about cryptic crickets and leaf-mimicking katydids and mantids, so they take a good close look, just to make sure. Many katydids respond to a monkey troupe's rummage-squad corporate searching methods by diving instantly for the only safe place – the ground. The author has on a number of occasions stood directly beneath a band of monkeys noisily ravaging the canopy above and seen leaf-like katydids desperately plopping to the ground around him.

If the katydids manage to survive the monkeys' daytime depredations, they still have to face a new threat after dark in the shape of another group of highly successful mammals, the bats. Sight-dependent hunters such as monkeys and birds have to be both perceptive and vigilant in order to spot prey in exposed yet cryptic daytime poses, or hidden in crevices and under leaves. At night katydids become active and the males forsake all attempts at concealment by taking up a prominent position on a leaf, in order to call and attract a mate. It is thought that bats trade on these calls, using them as homing-beacons, although this fails to explain how such large numbers of the silent, less publicly exposed females also end up in bat-droppings.

INVERTEBRATE ENEMIES

While vertebrate enemies are many and varied, and responsible for the evolution of most of the defensive devices mentioned in Chapter 5, there is also a wide spectrum of invertebrate organisms doing their bit to control the numbers of the orthopteroid insects. In fact, an attack by an invertebrate enemy may increase the risks of succumbing to a larger antagonist, as when the internal burden of a mass of parasitic fly larvae slow down a cricket's escape response, with the result that the whole lot ends up sliding down a bird's gullet. This may even decrease parasitism temporarily through selective cropping by vertebrate predators confronted with easier, parasite-ridden targets. However, internal parasites are only one problem, as there are numerous six-legged assailants capable of ending resistance in their prey just as effectively as any

Plate 98 Jumping spiders of the family Salticidae will take a wide variety of prey. This *Plexippus* species in Kenya has pounced upon a katydid larger than itself.

bird. These miniature killers may also be equipped with vastly superior senses for locating their victims in the first place.

The nearest invertebrate equivalent to a pouncing lizard is probably the capture-leap of a jumping spider. These spiders (family Salticidae) are blessed

Plate 99 Army ants are a serious threat to all rainforest insects, especially those which cannot fly or can only fly poorly. This cockroach in a rainforest in Trinidad tried to flee its assailants on foot but was soon brought down by weight of numbers and torn to pieces.

with excellent forward vision and will tackle anything of a size small enough to be overcome, including grasshoppers and katydids considerably larger than the spiders themselves. Wolf spiders (family Lycosidae) also hunt by day, while scorpions, sun spiders, whip spiders and vinegaroons are prowlers of the night, particularly in warm, dry areas. The threat posed by web-building spiders seems to vary according to the type of web. Sheet-webs furnished with abundant trip-lines, such as are found in the Agelenidae, must account for large numbers of grasshoppers, the more so as the webs are often constructed low down among grass and bushes. Once tangled on the trip-lines, the grasshopper flounders helplessly around for the split second needed for the spider to rush from its lair and administer a venomous bite. Spiders which construct aerial orb-webs seem to have less success, for the strong hind legs typical of grasshoppers often kick their way free of the ensnaring silk before the spider can get to grips with its catch. The larger the spider, however, and the stronger its silk, the higher the success rate. In Central America more than 10% of the prey captured in the webs of *Argiope argentata* was composed of katydids, grasshoppers and a few crickets. Although only representing one tenth by number, this group constituted more than 50% by weight of the spiders' food-supply, thus representing a valuable resource for the spiders.

Robberflies (family Asilidae), predatory beetles, such as tiger beetles, and assassin bugs may also pose a threat, but only in opportunistic terms as part of their generalized hunting procedures. This also applies to ants, although species which hunt in large swarms, such as the army ant *Eciton burchelli* of the American tropics, soon put paid to large numbers of all kinds of insects, particularly those which cannot escape the wide zone of attack by flying. Even jumping may not succeed in securing more than a temporary reprieve from such a prolific foe, for the most vigorous of leaps is usually insufficient to clear

Plate 100 When nymphs are at the stage of hatching from the eggs they are particularly vulnerable to predation by ants. These tiny *Azteca* species ants in a Costa Rican rainforest are plundering a batch of newly-emerging katydid nymphs. The eggs are of the flattened type placed in batches on leaves by certain tropical katydids.

Plate 101 Some orthopteroid insects prey on one another. This katydid is being devoured by the nymph of a *Parasphendale agrionina* mantis in tropical dry forest in Kenya.

the danger-area and may simply land the insect in even greater trouble, eventually to be dragged down and overpowered by sheer weight of numbers. Wingless insects, such as many cockroaches of the forest floor, numerous katydids and stick insects and the nymphal forms of all insects, are in a particularly desperate fix, although even good flyers may succumb if they do not react swiftly enough. In the midst of such slaughter the only hope of salvation may be to try and remain unnoticed. Despite their vast numbers, the ants cannot scour every bit of vegetation and the author has seen cockroaches survive in the face of enormous odds merely by retreating to the very tip of a long narrow leaf, the type of peninsular sanctuary most likely to be missed by the ants.

Ants pose a major but undirected menace. A much more specific threat stems from certain members of the orders Diptera and Hymenoptera, which are highly specialized to target only certain members of the orthopteroid group. Among the Diptera, the most important enemies are parasitic flies of the family Tachinidae. In South America, females of the tachinid fly *Euphasiopteryx depleta* are attracted to the calls of mole-crickets. These flies are remarkably skilled at following the sound-source, even though to human ears it has a confusing ventriloquial quality, and even intermittent calling does not hinder the fly's eventual arrival beside the caller. She deposits larvae near the burrow which attach themselves to mole-crickets nearby, especially to females turning up in response to the males' calling. In Texas, females of *Euphasiopteryx ochracea* orient to the song of male field crickets *Gryllus intiger*, depositing their larvae actually on the body of the cricket. The larvae of these flies bore into a host and consume it from the inside, eventually emerging again when they are ready to pupate.

The incidence of parasitism can be remarkably high. Rates of 90% have been reported for the nemestrinid fly parasite *Trichopsidea flavopilosus* in the desert locust *Schistocerca gregaria* in Africa. Such parasites are also remarkable for their ability to attack chemically defended grasshoppers which are shunned by vertebrate predators. The calliphorid fly *Blaesoxipha kaestneri* parasitizes the grasshopper *Poecilocerus pictus* which contains glycosides highly poisonous to vertebrates. *B. laticornis* females stalk walking locusts and insert their incubated eggs into the genital orifice of their victim. The larvae of *Sarcophaga destructor* are seemingly incapable of penetrating the tough cuticle of adult locusts, but are able to effect an entry when their intended host is soft and newly-moulted. *S. kellyi* exhibits breath-taking agility and ingenuity by chasing flying locusts and ovipositing on their wings. The females of *Stomorhina lunata* tag along with migrating locust swarms, waiting for the female locusts to find a suitable area in which to pause awhile and lay their eggs. The *Stomorhina* female sits next to the ovipositing locust, biding her time until the moment when the locust withdraws her ovipositor from the earth. The frothy protective plug filling the hole will still be soft, and it is in this that the fly now lays her own eggs. The fly's close shadowing of the female locusts is essential, for the parasite is unable to penetrate a mature plug whose froth has hardened. A single pod may provide board and lodging for several larvae, but even one occupant is so messy in its feeding and careless in its movements that the entire pod is usually destroyed.

Species of *Trichopsidea*, rather bee-like flies belonging to the family Nemestrinidae, are known to parasitize the Australian plague locust *Chortoicetes terminifera* as well as African *Locusta* and *Schistocerca* locusts. The first instar fly larva is a mobile creature called a planidium, whose job is to seek out and attach itself to a locust. It then tunnels into the host and feeds on the body tissues, although the presence of this parasite may not be so rapidly fatal or deleterious as in some other host-parasite relationships and a female grasshopper may even manage to lay a batch of eggs, despite her internal guest. The fully grown *Trichopsidea* larva bores its way out of the grasshopper and pupates in the ground. Here it may enter a resting or diapause state until stimulated to hatch by rain wetting the ground. This ensures close synchrony with the host grasshoppers, which will also be induced to breed after abundant rains.

Flies of the rather similar-looking family Bombyliidae also parasitize grasshoppers. The South African species of the genus *Systoechus* manage to locate and parasitize the buried egg-pods so carefully hidden by the female grasshoppers. The bee-fly larva makes a meal of a proportion of the grasshopper eggs before pupating in the ground. *S. somali* parasitizes the egg-pods of *Schistocerca gregaria* further north on the African continent. Some of the most amazing of all parasitic flies are various exclusively tropical species of *Stylogaster* belonging to the family Conopidae. The females hover around foraging swarms of army ants and dive into the vegetation to oviposit on insects fleeing in panic from the invading armies. Cockroaches are frequent targets, and the flies pounce on them and attach an egg. The egg itself is a remarkable instrument designed for firm atttachment to a mobile host, being secured by a combination of a barbed point and an inflatable sac which is pumped up by osmotic pressure.

The parasitic wasps in the order Hymenoptera comprise an enormous and extremely successful group with a vast range of different hosts. Larvae of the Evaniidae develop within the oothecae of cockroaches. Although the female wasp actually oviposits into a single egg, the larva soon runs out of food and leaves its original home to feed on neighbouring eggs. The eggs of many other orthopteroids are attacked by a variety of parasitic wasps, and there is sometimes a remarkably close relationship between parasite and host. When a female of the chalcid wasp *Rielia manticida* lands on a praying mantis, she jettisons her wings and attaches herself by her mandibles to the mantid's abdomen or the base of its wings. The mantis, which is usually a female, is destined to be exploited both as a source of food and as a courier direct to the wasp's intended host, the mantid's eggs. The wasp sucks the mantid's body-fluids and is on hand when the mantis makes its ootheca. It is essential that the wasp should be nearby at the start of egg-laying, for she cannot penetrate the tough outer envelope of a mature ootheca and must make the most of the period when the protective secretion is still soft and frothy. Other tiny parasitic wasps also avail themselves of mantids or grasshoppers as involuntary ferries to egg-laying sites, but do not feed on their living vehicle in the interim. The eggs of certain Australian stick insects are utilized by a tiny *Myrmecomimensis* parasitic wasp. The female wasp's wings are vestigial and she cannot fly, so she hastens on foot across the leaf-litter in her quest for phasmid eggs which have been scattered on the ground. She climbs up on to the egg, gnaws an aperture in its tough shell and then, after perhaps snacking briefly on some of the yolk, she backs into the hole and lays an egg. Each phasmid egg nurtures a single wasp larva and hatching of the adult parasites is often closely in synchrony with the commencement of egg-laying in the hosts.

Many kinds of solitary wasps belonging to the superfamily Sphecoidea are specialist predators on orthopteroids, usually restricting their attentions to certain groups (some of them even specialize on taking mantids!). They normally sting their victims first to induce a state of paralysis before dragging them to a nest-burrow, where the wasp's larva can munch away in peace on its living meal. The family Larridae specializes on crickets and mole-crickets. A female *Larra* attaches her egg to a mole-cricket which she has flushed from its underground burrow and disabled by stinging. The paralysis soon fades away and the mole-cricket goes back to its normal activities. However, it eventually succumbs to the steady gnawings of its unwelcome passenger, the ever-growing wasp larva. As mole-crickets are large, bulky insects, this habit of utilizing a still-active yet doomed host is the only practical method for the wasp. But other members of the family, e.g. *Liris* which utilize smaller crickets, first paralyse their prey and then drag it away, inert and helpless, to suffer lingering dismemberment by the wasp's larva in the privacy of its nest-burrow.

Numerous kinds of hunting wasps prey on grasshoppers or katydids. A *Sphex tomentosus* under observation by the author in Kenya brought a paralysed katydid to its burrow on average every 45 minutes. This is a remarkable feat considering that these particular katydids were fairly convincing mimics of leaves, yet the wasp's sensory apparatus is so highly developed that regular

Plate 102 Many solitary wasps are specialist hunters of certain kinds of orthopteroid insects. Species of *Liris* only take crickets. This female *Liris* is dragging a cricket which she has stung and paralysed. She will place it in her nest-burrow and lay an egg on it; its still-living body will be eaten by her larva. Photographed in tropical dry forest in Kenya.

detection and capture of such impressively cryptic quarry apparently presents no difficulties. It is likely that many more katydids may have escaped before the wasp could sting them, so the actual number discovered was probably higher, possibly much higher, than the observations at the nest would suggest. It is unlikely that any birds, notwithstanding their acute eyesight, would have been capable of such consistent detection of such a well-camouflaged insect. The ability to detect its prey using several senses in alliance, including scent (a sense of little use to a hunting bird), probably accounts for the wasp's success. The slim lightweight *Prionyx* wasp dragging a plump grasshopper to its grave over some open sun-baked terrain always looks as though she is scarcely up to such a laborious task. *P. parkeri* in the southwestern USA locates grasshoppers using visual cues. The prowling female attempts to take her prey by surprise,

Plate 103 A number of entomophagous fungi attack and kill insects. The corpse of this *Prionolopha serrata* grasshopper in a Peruvian rainforest is adorned with the club-shaped fruiting bodies of the fungus *Cordyceps locustiphila*.

but even if she successfully stings it, she may still encounter problems from the gob of fluid which the grasshopper often regurgitates. This is capable of gluing up the wasp's mouthparts or wings and may even kill it, as it seems to act as a contact poison.

The majority of larval blister beetles of the family Meloidae are parasites in grasshopper egg-pods. The adult beetles are often warningly coloured and may be conspicuous and common, especially in arid country. With her slim legs and rather small jaws the female South African *Mylabris* seems ill-equipped for her difficult task of excavating a pit an inch or so deep in hard soil. This receives a batch of 20 to 30 eggs and several batches are laid before she exhausts her reserves and dies. The eggs produce a highly mobile triungulin larva whose job is to locate a buried grasshopper egg-pod in which it will moult into a fat white grub, suited for an indolent life feeding off eggs in a secure environment.

Although parasitic and predatory invertebrates undoubtedly take a heavy toll, there remains another, perhaps rather unexpected and highly insidious enemy of such virulence that it is capable of virtually exterminating whole populations of orthopteroids in a very short time. This wholesale killer is the fungus *Entomophthora grylli*, an entomophagous species which mainly infects grasshoppers and crickets, although not all species are equally prone to infection and some, such as *Locusta migratoria*, seem to be immune. The fungus is at its most deadly during periods of warm wet weather, when the mortality in the more sensitive species of grasshoppers can reach 99% or more.

The final movements of the doomed grasshoppers are artificially manipulated to serve the interests of the invading fungus, and the hosts act in a strange and characteristic way induced by the infection. They clamber up to the topmost parts of grasses or bushes, wrap their two front pairs of legs around the stem and release their hold with the rear legs, which are splayed somewhat outwards and downwards. Death occurs soon after the assumption of this weird posture, and is followed within an hour or so by the emergence of white furry masses of fungal reproductive bodies from the various joints of the corpse. The reason for the elevated position adopted by the grasshoppers at their killer's behest is now obvious. The higher the insect, the better the spores can blow around and infect neighbouring insects, especially when the grasshoppers are in plague proportions and roosting in close proximity. Although the majority of congeners will succumb, there always seems to be a small percentage who appear to be completely resistant to infection, even when covered in a blizzard of infective particles. A number of other fungi also infect orthopteroids, but none with such potentially devastating effects as this one.

Chapter 8

Interaction with the Human World

Grasshoppers have been in the news for a long time. Even in biblical times a plague of locusts was front-page news. This is hardly surprising. A single swarm of Africa's most devastating plague species, the desert locust *Schistocerca gregaria*, may fan out over hundreds of square miles and consist of 50 billion individuals with an all-up weight exceeding 70,000 tons (71,100 tonnes). Densities may attain in the region of 200 million locusts per square mile (500 million per sq km). The amount of food devoured in a single day by such a swarm is prodigious, equalling the daily consumption by the citizens of New York, London, Los Angeles and Paris. In 1794 a swarm that spread over 2000 square miles (5178 sq km) succumbed to one of the locust swarm's deadliest enemies – the wind – and was blown out to sea from the coast of South Africa. Within days the tide had swept ashore a wall of corpses 4 ft (1.2 m) deep and extending for 50 miles (80 km) along the coast. Even this is not the record, for the largest grasshopper swarm of all time was recorded from Oregon and California in 1949. Its zone of destruction extended across 3000 square miles (7770 sq km) and the damage done was enormous. It is no small wonder, then, that peasant farmers in a dry and hostile land relying on a meagre, hard-won crop for their family's survival over the coming year have always viewed the arrival of the ravening hordes with a mixture of fear and awe. No wonder either that the Bible itself reported plagues of locusts as one of the greatest forms of pestilence to be vented against the land and its peoples.

A locust swarm is a remarkable affair, behaving almost like a single, multi-faceted organism, each part of which responds to controlling factors as the parts of a body react to commands from the brain. Yet the origins of a swarm seem so harmless, with little portent of the threat to come. During normal dry weather desert locusts exist as sparse individuals gleaning a poor living from the almost barren land. The nymphs are drab creatures, well camouflaged in their dry surroundings, little given to associating with their fellows, although the age-old union of male and female still takes place when the mating drive brings the adults together. Even in such hostile circumstances the females still respond to the urge to lay their eggs in the hard-baked ground.

Yet far away beyond the horizon, momentous events are shaping up which will transform both this barren scene and its attendant locusts. A low-pressure area is forming over the North African desert, the prelude to rain, which sweeps in and deluges the whole region, perhaps the first proper downpour for several years. The change wrought in the landscape is dramatic. A green carpet of soft young vegetation cloaks the once arid ground. The locusts are enjoying a feast, females are ovipositing all over the place and the hatching nymphs thrive and grow rapidly. As their numbers quickly mount, the nymphs lose their unsociable habits and actually begin to court the company of their fellows. Finding one another is now made far easier, for instead of being drab and camouflaged, this new generation of nymphs is composed of conspicuous insects flamboyantly attired in black and yellow.

Small groups of nymphs soon coalesce to form large, colourful bands of hoppers which march onwards across the land in a bright, living tide. Soon after the hoppers have become adult, the food-supply is exhausted. The response is spectacular, but tragic for the human inhabitants in their path. They take to the air in their millions, often following the prevailing winds of a cold front which could lead them to areas of fresh pasture. An oasis of green is the magnet which sends them spiralling earthwards in a vast demolition squad which strips much of the vegetation bare. Locusts above all relish grasses, which spells disaster for peasant farmers' crops of millet, sorghum, maize or wheat. If the area contains sufficient food the locusts will again stop to breed. Often they cover huge distances in their continuing quest for greenery, until they finally run out of luck and billions of them die of starvation, leaving a trail of havoc in their wake. Those which do manage to lay their eggs in some parched habitat will give rise to drably coloured nymphs with unsocial habits. The wheel has turned full circle.

Many types of grasshoppers are called locusts, but the name should really be reserved for those species which occur in two distinctive phases, solitary and gregarious. The dividing line between the two is not always sharply defined and under certain circumstances the transition from one to the other can be interrupted or even reversed by changes in environmental factors. When fully developed, the two forms are so different in their colours, habits, physiology and ecology that in the past they have been regarded as belonging to two completely different species.

The Sahel region of North Africa is the area now most likely to experience the ravages of swarms, mainly of the migratory locust *Locusta migratoria* and the desert locust. Other continents suffer the effects of various other locusts, as well as plague grasshoppers, which do not have different phases but still cause considerable problems to agriculture when conditions permit. In Thailand, *Patanga succincta* often does considerable damage to maize and has to be controlled by insecticides, although a natural control in the form of an entomophagous fungus sometimes does the job instead. These fungi have been under consideration for use in biological control of grasshopper pests, but at present there are drawbacks due to the fungus's requirement for an extended period of moist warm weather to be effective as a control. Really massive, devastating swarms of biblical proportions are now less common as a result of careful monitoring of the weather conditions and locust numbers in the areas most at risk, such as the Sahel. The idea is to nip the problem in the bud by directing the main control measures at the vulnerable and crowded hopper stages, before the highly mobile aerial phase is initiated.

While locusts are the most spectacular and newsworthy orthopteroid pests, a number of other species interfere with the ordered state of man's affairs in a variety of ways. The colourful variegated grasshopper *Zonocerus variegatus* is a serious pest of cultivated cassava crops in West Africa, while its equally beautiful relative *Z. elegans* damages a wide assortment of crops in eastern and southern African. In North America the black-horned tree cricket *Oecanthus nigricornis* damages twigs of peaches, apples, grapes and roses by laying its eggs in rows inside the pith. The two mole-crickets *Scapteriscus vicinus* and *S. acletus* are pests in the warmer parts of the southeastern United States. They were accidentally introduced sometime last century or earlier this century and have become widespread. Their burrowing activities and feeding habits can cause severe damage to lawns, golf-courses, pastures and growing crops.

A few species of stick insect can occur in outbreaks of sufficient magnitude to be of economic importance. *Diapheromera femorata* is the culprit in North American forests, while *Graeffa crouani* may wreak havoc in coconut plantations in the South Pacific. In Australia a number of stick insects are major pests of Eucalyptus trees. The three species involved are *Didymuria violascens, Carausius tessulata* and *Podacanthus wilkinsoni*. When an outbreak occurs, large numbers of birds soon move into the area and gorge on the stick insects, but with little apparent effect on their numbers – there are simply too many of them. This may even have the opposite effect on the total outbreak, for by concentrating on the core areas where their prey can be picked off the trees without really trying, the birds let the phasmids in the surrounding areas off the hook and allow them to build up their numbers unhindered. This may lead to wavefronts of infestation surging outwards from the central area in a kind of chain-reaction. A far more catastrophic 'crash' in the population happens when the crowds of phasmids, with their hearty appetites, finally mop up the last few morsels of living vegetation and die of starvation.

The consequences of a serious outbreak can be quite ruinous. In 1963 *D. violascens* managed to strip bare 650 square miles (1683 sq km) of Eucalyptus forest in southern Australia. The alpine ash *Eucalyptus delegatensis* tends to suffer a high mortality when completely defoliated and may die out over quite a wide area. Even when the trees survive, their vigour is severely reduced, as they devote their already diminished resources to building a fresh crown. This reduces the amount of material added to the trunk, as well as potentially permitting soil erosion of the exposed ground and the vigorous growth of grasses which could inhibit the establishment of seedling trees. The complete cycle of forest renewal may therefore be interrupted or even blocked. Repeated defoliations can spell disaster for the trees, to the extent that 83% of mountain ash *Eucalyptus regnans* have been recorded as dead within two years of suffering a second bout of defoliation.

In Michigan, ten to fifteen years of repeated assaults by *Diapheromera femorata* may wipe out more than half the black oaks in the area of attack, although associated trees such as aspens, maples, pines and even white oaks remain unmolested. Such selective feeding by the phasmid may eventually lead to a radical alteration of the forest's makeup by exterminating the black oak. In the Pacific, *Graeffa crouani* devours all the softer tissue of the coconut's fronds, leaving only the parts too tough to manage, such as the mid-rib and branching veins. Older palms suffer the worst and may die as a result, although loss of that year's crop is the most frequent outcome, and subsequent crops are reduced until a full head of fronds is again in place. Attempts at tackling outbreaks of pest phasmids using biological control methods have so far met with little success and chemical control seems to be the only practical proposition at present.

Plate 104 Two adult Australian cockroaches *Periplaneta australasiae* gorge on a sweet bun. This cosmopolitan pest is one of several which man has spread throughout the world. They foul as much food as they eat and are very difficult to eradicate.

Their habit of eating plants which are of economic importance to mankind is not the only way in which orthopteroid insects may make a nuisance of themselves. Stored products are also at risk from those insect equivalents of rats and mice, the cosmopolitan Australian cockroach *Periplaneta australiae*, the American cockroach *P. americana* and the oriental cockroach *Blatta orientalis*. Like rats and mice, these three species have been involuntarily distributed all over the world by man's activities and will eat a wide variety of foodstuffs. Cockroaches are particulary catholic in their tastes and as well as tucking into the more usual comestibles such as bread or biscuits, they will happily chew away at objects displaying less obvious delectability such as papers and book-bindings. Unfortunately, it is not just the physical loss of the food which is important, but the amount which is damaged or fouled by contact with a cockroach which may spend the day hiding away in a sewer or toilet. Their feet, mouthparts and droppings are effective vectors for transmitting diseases and cockroaches have been noted carrying such deadly organisms as the poliomyelitis virus and the *Salmonella* bacteria which can cause such devastating and potentially lethal outbreaks of food poisoning. Unfortunately, cockroaches are not just associated with Third World conditions or old dilapidated houses, for they may infest modern buildings such as hospitals and high-class hotels, where the central heating and other ducting ramifying throughout the building provide warm and convenient homes and highways for the nocturnal wanderings of their six-legged denizens.

Cockroaches share with rats and mice two other characteristics which have similarly inauspicious implications for mankind. They are difficult to eradicate, for they soon learn to avoid poison baits once they have suffered an unpleasant but not fatal experience. They also reproduce rapidly and are, for insects, extremely long-lived. The American cockroach comes out top on both counts, for females may attain a grand old age of four years, time well spent in churning out a cockroach production-line of over 1000 eggs. With these kinds of assets at their disposal, cockroaches, rats and mice are all equally difficult to control and all equally likely to enter the twenty-first century with every prospect of continued success in eluding man's best efforts at eradication. The greatest promise for success lies in the control of isolated populations, such as occur on islands. Parasitic wasps of several species have been introduced to the Cook Islands to help control the American cockroach, with at least an initial measure of success.

The story of man's relationship with the orthopteroid insects is not, however, solely a negative tale of woe and damage. There is a more positive side, as well as a further negative aspect, this time concerning man's adverse effects on the insects themselves. Firstly, in a positive vein, a number of grasshoppers and crickets have been considered for use as biological control agents against other types of pests. A grasshopper showed some promise against a water fern which has become a major pest after accidental introductions to many tropical waterways, although a weevil eventually proved to be the solution to this particular problem. The cricket *Metioche vittaticollis* is an important predator of the eggs of various rice pests in southeast Asia, including defoliators, stemborers and leaf-folders, insects which are notoriously resistant to effective control by chemical means. This cricket may therefore prove of use in biological control, although the practical applications would be something for future development.

Perhaps the most surprising connection between man and the orthopteroids is the use of a number of these insects as human food. Various types of grasshoppers, katydids or crickets have figured on the menu in just about every country for which details of dietary intake have been recorded. In Africa the edible bush cricket or 'nsenene' *Ruspolia differens* (Tettigoniidae) is eagerly sought after by the local populace. This insect often migrates at night in huge swarms which may be attracted to lights in towns. Here they can be swept up in their thousands by excited crowds of locals grateful for the bounty which has so conveniently landed up on their doorstep. In Thailand there are records of considerable enterprise being shown by the local officials in a province severely afflicted by a plague of grasshoppers. After steps to exert control had failed dismally, the locally citizens were exhorted to turn their efforts towards collecting the insects instead, with the result that more than 10 tons of grasshoppers were soon in the bag. Shops and restaurants in the area quickly began featuring crispy-fried grasshoppers, which apparently went down a treat with the customers. Such was the burgeoning demand that the area soon began exporting its delicacies, which rapidly won a place as a favourite dish, savoured not only for its flavour but for the ease of preparation – just pluck off the crunchy bits such as wings, legs, heads and tails, give them a quick swill and then straight into a pan of sizzling oil. In Oaxaca State in Mexico, *Sphenarium* grasshoppers are sold in the markets after having been seasoned with onion, garlic and chilli powder, then boiled, dried in the sun and finally fried.

Western readers revolted at the thought of tucking into a plate of stir-fried grasshoppers might bear in mind that the consumption of raw, still-living oysters is considered a delicacy by many of the most aristocratic of western citizens, while snails and marine shellfish are eaten with considerable relish by sophisticated gourmets. In North Africa, the local people have a history of mitigating at least part of the effects of a locust attack on their crops by turning on the locusts and eating them instead. A report dated 1891 speaks of 60 camel loads (approximately 20,000 lbs or 9000 kilos) of the desert locust being collected daily in one area alone. The poor people regarded them as a valuable addition to their diet. They are apparently very tasty when lightly fried in oil (after first removing the legs and wings, of course) but can easily be preserved for later consumption. This was done by cooking them in salt water and then drying them in the sun. A large swarm would not only easily supply the immediate and future needs of the local populace, but also left enough over for trading in the markets.

Grasshoppers have even played a vital role in deciding the outcome of human conflicts. During a great battle many years ago in Mexico, the local Anahuac Indians had to retreat to the safety of the summit of a nearby mountain. A plentiful water-supply coupled with an abundance of edible grasshoppers enabled them to hold out without starving and saved the day. Their hilltop asylum stands near the centre of present-day Mexico City and the role played by the grasshoppers is commemorated in its name of Chapultepec or 'Grasshopper Mountain'.

Making the best use of pests by adopting the pragmatic philosophy that if something beats you to your crops, then eat the something which got there first is also practised in the Far East and in central Africa. The large pest cricket *Brachytrupes portentosus* is widely offered for sale in markets in Burma, Malaysia and Indonesia, destined for human consumption. The thrifty native peoples exploit the crop-guzzling *Brachytrupes membranaceus* in Uganda and neighbouring parts of East Africa in a similar way, although in none of these instances is there a planned, regulated effort to make the best and most efficient use of these pests as dietary supplements.

A project in Nepal is dedicated to improving efficiency by perfecting light-trap harvesting methods for *B. portentosus*. The eventual use for the catch is as a high-protein feed for poultry, replacing the current, expensively imported fish-meal. This neatly shortcircuits any human objections to eating crickets directly and is highly efficient, as dried insects have already proved comparable to or better than soyabean meal as a feed for rearing chicks. Another contendor for this use is the Mormon cricket *Anabrus simplex* (Tettigoniidae), which has for many years enjoyed the status of an expensive pest in the south-western part of the United States. The crickets form large, easily accessible bands, which might make collecting them for poultry food a more sensible way of tackling the problem than spraying them with costly and environmentally questionable chemicals. When hordes of these crickets wiped out the Mormons' crops in their newly-established home in Utah, the people were saved by an enormous flock of gulls which descended from heaven (or so it seemed to the besieged Mormons) and gorged themselves on the swarming insects. To this day a statue honouring these winged saviours stands in Salt Lake City.

By taking full advantage of the ability of certain insects to thrive on a diet of what we would consider to be purely waste material, the implications for poultry production are even more interesting. The house cricket *Acheta domesticus* will munch away on just about anything which is remotely edible. If chicken manure is mixed with ground corn, the crickets will not only take happily to such a seemingly unappetizing diet, but will actually thrive on it and convert the waste material to useful protein at an extremely inexpensive rate. This excellent rate of return could even be improved by dispensing altogether with the corn and replacing it with agricultural wastes from the pineapple, banana or similar industries.

A rather less productive way of using crickets for monetary gain has long been enjoyed in Hong Kong, where money is wagered on the outcome in the popular sport of cricket-fighting. The most valiant of these mini-pugilists are the so-called 'human-head' crickets whose normal domicile is the interior of human skulls in graveyards. With seven holes serving as escape-hatches in each skull, these crickets are, not surprisingly, difficult to catch, as they are inclined to dart elusively from one part of the skull to another via the holes. The only sure method is to dip the whole skull in water, a drastic but sure way of driving the crickets out.

The singing of crickets has long been valued by those with an appreciative ear for such a monotonous output. In former times, special 'singing-cages' were constructed to hold the diminutive virtuosos, but with the remorseless advance of transistor-induced sound-effects, the days of such simple rustic entertainment are surely numbered, even in the most remote of rural backwaters.

Despite repeated depredations by the owners of fighting crickets over the years, there presumably still remains a rich supply of livestock to be found in the local graveyards in Hong Kong. Such is unfortunately not the case with a number of highly interesting species which have been almost exterminated by man's malign influence. The wholesale destruction of the tropical rainforests which is now taking place could soon bring hundreds if not thousands of species of orthopteroids to extinction or to its brink. This has already happened on a more minor scale, where species which are adapted to life in a relatively secure, predator-free habitat have suddenly been faced with a major new threat introduced by man. This kind of sad scenario has mainly taken place, at least so far, in New Zealand and its associated islands, where the local fauna and flora are very vulnerable to competition from introduced species.

The Lord Howe Island phasmid *Dryococelus australis* is one of a number of large New Zealand orthopteroids which have been almost eliminated by those most adaptable and destructive of man's fellow-travellers – rats. This large, clumsy, flightless phasmid made an easy and very abundant target for rats which were accidentally introduced to its island fastness. It may still maintain a precarious foothold on the nearby and very isolated Ball's Pyramid, but even a temporary incursion by rats on this last stronghold would soon snuff out the final population of the phasmid, even if a population does indeed survive there. The *Deinacrida* giant wetas (Stenopelmatidae) on mainland New Zealand have also suffered horribly from the depredations of rats. The three species of giant wetas hold the record as the heaviest of all insects – a mature specimen may tip the scales at $2\frac{1}{2}$ oz (70 g) which is nearly three times the average for a house mouse. Being wingless and incapable even of jumping to escape their enemies, the plump, shiny, coffee-coloured wetas have been stuck in an evolutionary full-stop for millions of years, changing not at all during that time. Each of the four species now survives only on offshore islands where introduced rats, pigs or cats have at times posed threats and could do so again in the future if care is not taken. They are all strictly protected by law and efforts have been made to establish populations of the more threatened species on new island homes which are free of rats and likely to remain so.

Glossary

Aposematic Warningly coloured.
Australian Zoogeographical region including Australia, Tasmania, New Zealand, New Guinea and numerous islands in the Pacific.
Cerci A pair of appendages at the end of the abdomen.
Crepitation Sound made by certain displaying male grasshoppers by cracking together of the hind wings.
Crypsis Being coloured and/or shaped to blend into the background.
Diapause A period of suspended animation in insects during unfavourable climatic conditions.
Elytra The first pair of wings modified to form wing cases, which protect the second pair of membranous flying wings.
Endophytic Laying eggs inside a plant.
Entomophagous Used of, for example, fungi, which feed on the body of an insect; literally 'insect eating'.
Epiphytic Laying eggs on the outside of a plant.
Fire melanism Assuming a dark coloration to blend into the blackened vegetation remaining after a fire.
Frass Droppings and other solid waste from an insect.
Instar A stage in the life cycle of an insect.
Leks Gatherings of male animals whose behaviour then attracts the females for the purpose of courtship and mating.
Mopane A type of African tree.
Myrmecophilous Being attracted to, or liking, ants.
Neotropical The zoogeographic region which extends from central Mexico to the southern tip of South America.
New World From the Americas.
Ocelli Simple eyes, used not for vision but to detect light and dark.
Old World From Africa and Asia.
Ootheca Egg case.
Oriental The zoogeographical area of Asia extending from 30 degrees N down to 30 degrees S of latitude.
Ovipary Reproducing by laying eggs.
Ovipositor Egg-laying tube on the female.
Ovovivipary Retention of the eggs in the body until they are about to hatch.
Palps Head appendages concerned with the sense of touch and taste.
Pheromone An airborne chemical messenger, usually for the attraction of members of the opposite sex.
Pronotum A backwards extension of the dorsal surface of the prothorax (q.v.).
Prothorax The first of the three segments of the insect thorax.
Sexual dichromatism The two sexes being differently coloured.
Sexual dimorphism The two sexes having distinctly different physical appearances.

Speciation The evolution of distinct species.
Spermatophore A packet of sperms released by the male to be picked up by the female.
Spermatophylax That part of the spermatophore (q.v.) which acts as a special food source for the female.
Stridulation The production of sound to attract a mate.
Substrate The surface on which an insect lives.
Tegmen (pl. tegmina) The forewings of an orthopteran.
Tremulation The male insect attracting a female by vibrating his body against, for example, a leaf, which then amplifies the sound produced.
Troilism Competition between males for a single female.
Tympanum The ear-drum.
Tympanal aperture The opening into the ear-drum.
Vermiform Worm-like.
Vivipary The production of living young rather than eggs.

Index

Numbers in *italic* refer to black and white illustrations.
Numbers in **bold** refer to colour plates.